Nexus Network Journal

Canons of Form-Making
In Honour of Andrea Palladio 1508-2008

Stephen R. Wassell and Kim Williams, editors

Volume 10, Number 2
Autumn 2008

KIM WILLIAMS BOOKS

Nexus Network Journal
Vol. 10
No. 2
Pp. 207-380
ISSN 1590-5896

CONTENTS

Perhaps no other single architect has had an impact on the face of Western architecture as has Andrea Palladio (1508-1580). The characteristic forms used in his villas, basilicas and palazzos were adopted and adapted for widespread use first in England and then in the United States and are now as representative of those lands as they are of the Veneto which saw their birth. But perhaps an even greater legacy of Palladio is his *Four Books of Architecture*, one of the first treatises on architecture that was richly illustrated and intended for a readership of architects and builders rather than intellectuals. Thanks to the clarity and scope of the *Four Books*, they can still be studied with profit today. What makes the *Four Books* of enduring interest is that Palladio set forth his canons of architecture, that is, the rules he used to create his architectural forms, from the details and proportions of the orders to the layout of floor plans for various building types. These rules, rather than remaining specific to a single building type at a unique moment in time, have been studied and abstracted and reapplied to find new, fresh applications. This is the aspect we most wish to honor with this special issue of the *Nexus Network Journal*, entitled "Canons of Form-Making," dedicated to the quincentenary of Palladio's birth.

The issue opens with Stephen R. Wassell's "Andrea Palladio (1508-1580)". This brief biography was originally written for "The Year of Palladio" website of the Institute for Classical Architecture and Classical America (http://www.classicist.org/resources/year-of-palladio/). The editors therefore wish to thank the ICA&CA for permitting us to publish the biography in this special issue of the *NNJ*. For inclusion in this issue, the endnotes and bibliography have been expanded from the original version to highlight numerous publications concerning relationships between architecture and mathematics in Palladio's oeuvre. Of course, we can now add to this bibliography the following three articles devoted to Palladio!

Lionel March's "Palladio, Pythagoreanism and Renaissance Mathematics" is an analysis of the very first building presented in the crucial second book of Palladio's treatise, the Palazzo Antonini. In this article March applies his expert knowledge of the types of mathematics (including what we would now call numerology) that were important to Renaissance scholars and practitioners, in order to gain a better understanding of Palladio's design methodologies and canons.

Coincidentally, Buthayna Eilouti's "A Formal Language for Palladian Palazzo Façades Represented by a String Recognition Device" also analyzes Palazzo Antonini, as well as the rest of the nine designs Palladio includes in book II, chapter 3. But while March focuses on mathematics known to Palladio, Eilouti analyzes the nine façades using modern mathematics related to computer science, namely regular languages and finite state automata.

In "A Perspective Analysis of the Proportions of Palladio's Villa Rotonda: Making the Invisible Visible", Tomás Salgado-Garcia examines the very familiar Villa Rotonda from the viewpoint of perspective to show that Palladio's proportions are not buried as abstract concepts visible only on the drawing board but reveal themselves visually in the building.

Canons of architecture existed of course in antiquity, long before Palladio set his own down in writing, but in the face of the lack of written documentation, they are tantalizingly hidden. New mathematical tools can help researchers uncover those canons. In "The Doric Order as Fractal", Carl Bovill uses the technique of iterated function system (IFS) to analyze the Doric temple, and shows that the characteristics of self-similarity and self-affinity that result from it indicate that the Greeks based their canons of architecture on what they observed in nature.

Michael Duddy also examines the Doric order in "Roaming Point Perspective: A Dynamic Interpretation of the Visual Refinements of the Greek Doric Temple". He examines the very subtle canons for making corrections to structures or elements so that they look like they should ideally, and presents a new theory that holds that these corrections were efficacious as the observer changed position, not just from a static, "ideal" viewing point.

Roberto Castiglia and Marco Giorgio Bevilacqua take us to Albania to discover canons of Islamic architecture. In "The Turkish Baths at Elbasan: Architecture, Geometry and Well-Being", Castiglia and Bevilacqua report on the results of a survey campaign coordinated by the University of Pisa, which has provided precise information of the geometric rules that underlie the complex plans and the domes and vaults of the *hamman*.

All of these studies show that mathematics is a powerful tool for the architect, both during the original design process and in the after-the-fact study of existing monuments. Anat David-Artman believes that mathematics is not just powerful but vital. In "Mathematics as a Vital Force in Architecture", the work of biologist Hans Dreisch is used to forge an analogy between the principle of individuation and equipotential in life forms and in architecture.

Regarding didactics, in "The Use of Linear Fractional Transformations to Produce Building Plans" Christopher Stone gives a step-by-step explanation of how mappings on a complex plane can be used to generate floor plans.

In this issue's Geometer's Angle column, geometer Rachel Fletcher turns her attention once more to Palladio, as she has done twice before for Nexus readers and conference participants. In part three of her examination of Dynamic Root Rectangles, she explains the properties of root-three rectangles and applies them to the plans of Palazzo della Torre in Verona and Villa Mocenigo, Marocco in Treviso.

This issue concludes with two book reviews. Kim Williams reviews *Andrea Palladio: The Villa Cornaro in Piombino Dese* edited by Branko Mitrović and Stephen R. Wassell. Sylvie Duvernoy reviews *Architettura e Musica nella Venezia del Rinascimento* edited by Deborah Howard and Laura Moretti.

This is not only an anniversary year for Palladio, but one for the *Nexus* community as well. This is the tenth year of publication of the *Nexus Network Journal,* and it was ten years ago that the post-Nexus 1998 conference workshop was a tour of Palladio's villas. We co-editors, who led the 1998 workshop, learned much on that tour and as you can see, are still exploring the depths of Palladio's oeuvre ten years on!

Happy birthday, Andrea Palladio!

Stephen R. Wassell

Department of Mathematics
and Computer Science
Sweet Briar College
Sweet Briar, VA 24595 USA
wassell@sbc.edu

Keywords: Andrea Palladio,
Renaissance architecture, ratio
and proportion, harmonic
proportions, geometry, design
theory, classical architecture

Research

Andrea Palladio (1508-1580)

Abstract. A brief description of Palladio's life and works. The focus is on the evolution of his design methodology, including the growing importance of proportion to his approach. Selected mathematical details are cited in the endnotes, and the list of references includes many publications focused on the relationships between architecture and mathematics in Palladio's designs.

All over the western world, hundreds of thousands of houses, churches and public buildings with symmetrical fronts and applied half-columns topped by a pediment descend from the designs of Andrea Palladio. He is the most imitated architect in history, and his influence on the development of English and American architecture probably has been greater that that of all other Renaissance architects combined.

[Ackerman 1966: 19]

Half a millennium ago in Padua, a prominent city in the Veneto region of Italy, decades before the future architect would adopt the moniker Palladio, Andrea di Pietro dalla Gondola (1508-1580) was born into a family of modest means, his father Pietro being a mill worker. In 1521 Palladio was apprenticed to a Paduan stonemason, Bartolomeo Cavezza, but he broke away from Cavezza in 1524 and moved to Vicenza. To this day his adopted city celebrates Palladio as its most famous citizen.

Fig. 1. Villa Godi (photograph by the author)

Here Palladio joined the Pedemuro bottega, a workshop of stonemasons that enjoyed a steady stream of sculptural and architectural commissions, due in no small part to Vicenza's wealth. The workshop's consistent use of "protoclassical elements" with a "rare and unusual virtuosity in the range of artisan skills" undoubtedly influenced Palladio, although the workshop's classical "references have the effect of quotations embedded in vernacular architecture."[1] Palladio inherited his first architectural commission, Villa Godi (begun c. 1537, fig. 1), from Pedemuro at about the same time that he left the bottega.

During these formative years, Palladio developed the origins of his classical architectural vocabulary, deriving elements from many skilled architects, scholars, and practitioners in the surrounding region. These included the Pedemuro master Giovanni di Giacomo da Porlezza; the prominent Paduan patron Alvise Cornaro and his architectural circle, most notably Giovanni Maria Falconetto, Michele Sanmicheli, Jacopo Sansovino, Giulio Romano, and Sebastiano Serlio, whose nascent treatise on architecture was available to Palladio, books III and IV having been published by 1540. Palladio's drawings of classical elements such as capitals and entablatures from this time period show his desire to exercise his growing vocabulary, and it is telling that he later modified a number of these drawings after seeing the original buildings with his own eyes.

What was clearly missing in his early years was the first-hand knowledge of the Roman architectural sources upon which any respectable classical language must be based. Instrumental in bridging this gap was Giangiorgio Trissino, an aristocrat, writer, and humanist, who recognized Palladio's tremendous potential and facilitated his first trip to Rome in 1541. Palladio returned to Rome a number of times, with and without Trissino, where he meticulously researched, drew, and recorded copious amounts of architectural information, from ancient Roman sources to Renaissance masters such as Bramante, Raphael, and Michelangelo, from intricate details to overall plans and elevations.[2] Trissino also had formed an academy of sorts at his estate Villa Trissino in Cricoli (near Vicenza), on which Palladio had worked during his Pedemuro years. Here Trissino provided a humanistic education to promising scholars. He became a mentor to Palladio in the late 1530s and probably created the young architect's pseudonym. Through his study of the classics with Trissino, especially the architectural treatise of Vitruvius, and with the first-hand knowledge of Rome he acquired over several years, Palladio transformed his design approach substantially.

Let us first consider his earlier designs, during the 1540s, before the full extent of his education had taken force. Palladio makes scant use of the orders and other classical elements in many of his early villas. Instead he exhibits an innate interest in geometry as design medium, using simple forms such as the circle and semicircle to adorn his early façades, e.g., Villa Valmarana at Vigardolo (begun 1541, fig. 2) and Villa Poiana (begun c. 1548, fig. 3).

Symmetry is a constant stabilizing force early on and remains so throughout his career. Palladio's façade motif comprised of a three bay arcade surmounted by a pediment can inherently be viewed as a formal abstraction, a template from his toolkit – one that he realized at least four times, twice with rustication, once with orders, and once with minimal treatment; cf. Villa Pisani at Bagnolo (begun c. 1542-45), Villa Caldogno (begun c. 1545), Villa Gazzotti (begun 1541-42), and Villa Saraceno (begun c. 1545-8).

Palladio makes use of the orders, on palazzos and a small number of villas, to convey the importance of the owners, yet a major formal role of the classical elements is to regulate the steady rhythm of the principal façade (see, e.g., Palazzo Civena [begun c. 1540], Palazzo Thiene [begun c. 1542-6], and Palazzo Iseppo Porto [begun c. 1549]). In his interiors Palladio demonstrates a natural genius for shaping space while addressing programmatic concerns; of particular note is his use of vaults, individually and in combination, which, whether frescoed or left monochrome, read beautifully in their form, geometry, and structural grace.

Fig. 2. Villa Valmarana (photograph by the author)

Fig. 3. Villa Poiana (photograph by the author)

Towards the end of the 1540s and into the 1550s, as Palladio's classical vocabulary developed, there emerged a more advanced approach to his praxis. While still utilizing his mastery of geometry and creative ingenuity, Palladio fully embraced the use of the classical orders, which he integrated in more sophisticated ways. The previous uniform rhythm of pilasters or columns along the entire front façade becomes richer through the use of column groupings and varying intercolumniations, or it gives way completely to a central pedimented zone containing the orders. Entablatures become more articulated and prominent. Orders of different scales are combined. Intercolumniations are reduced in order to better conform to Vitruvian specifications. Buildings from this time period include the famous Basilica loggias (begun 1549) that adorn the main piazza of Vicenza and helped establish Palladio as the preeminent architect of the Veneto, Palazzo Chiericati (begun c. 1551), Villa Pisani at Montagnana (begun c. 1552), and Villa Cornaro (begun c. 1552, fig. 4).

In combination with his more sophisticated use of the orders, Palladio also takes strides in unifying the various components of plan, elevation, and section, an approach he would later describe in a sort of general maxim:

> Beauty will derive from a graceful shape and the relationship of the whole to the parts, and of the parts among themselves and to the whole, because buildings must appear to be like complete and well-defined bodies, of which one member matches another and all the members are necessary for what is required.[3]

Fig. 4. Villa Cornaro (photograph by the author)

This unified approach, which Palladio further refined during the 1550s and 1560s, was due in large part to the influence of his highly accomplished patron, Daniele Barbaro, a Venetian patrician, scholar, and humanist, who started working with Palladio soon after Trissino died in 1550. Palladio provided illustrations for Barbaro's translation and commentary of Vitruvius, published in 1556. Barbaro's commentaries are quite involved on topics concerning proportions, from the theory of the orders in books III and IV, to music in book V, to room ratios in book VI.

Palladio undoubtedly heightened his command of classical Roman architectural theory, as well as ancient Greek arithmetic and geometry, through his relationship with Barbaro.[4]

Villas from Palladio's later period include Villa Barbaro (begun c. 1556, fig. 5), Villa Malcontenta (begun c. 1558, fig. 6), Villa Emo (begun c. 1560, fig. 7), and the design widely considered to be his masterpiece, Villa Rotonda (begun c. 1566, fig. 8). His approach to design in this period has been the inspiration for much analysis. In "The Mathematics of the Ideal Villa" Colin Rowe compares Villa Malcontenta to Le Corbusier's Villa Stein, focusing mainly on proportional considerations [Rowe 1982: 1–27]. Villa Emo is one that most easily fits into Rudolf Wittkower's "harmonic proportions" formulation of Palladio's design theory, which has generated much interest and scrutiny since its publication about fifty years ago.[5] It is true that such proportional analyses can be used to argue that the beauty of Palladio's architecture is not necessarily tied to his use of the orders, and that architecture devoid of ornamentation can still delight the eye if aesthetically pleasing proportions are incorporated.[6] It is impossible to deny, however, that classical architecture has stood the test of time, and to this day classical designers look to the timeless beauty of Palladio's oeuvre for guidance and inspiration.[7]

Fig. 5. Villa Barbaro (photograph by the author)

Fig. 6. Villa Malcontenta (photograph by the author)

Fig. 7. Villa Emo (photograph by the author)

Fig. 8. Villa Rotonda (photograph by the author)

The death of Sansovino in 1570 left open the position of proto of the procurators of San Marco, i.e., the primary architect of Venice. Although Palladio never officially filled this position, during the 1560s and 1570s he succeeded in landing major commissions in the capital city, aided by his relationship with the Barbaro brothers, Daniele, who also died in 1570, and Marc'Antonio. Rather than adapt to Venetian styles, Palladio applied the architectural lexicon that he had derived from Roman sources and had mastered through years of study and practice. Palladio designed the front façade (commissioned 1562) of San Francesco della Vigna in a style that solves the problem of unifying the height of the central nave with the lower sides, by overlapping a classical temple front spanning the whole elevation with a colossal order fronting the nave. Variations on this same theme are seen on two churches in Venice fully designed by Palladio (albeit not completed until after his death), San Giorgio Maggiore (church begun 1566, fig. 9) and Il Redentore (begun 1577). The interiors of these two may seem a bit austere to some observers, since the only ornamentation is that which is implicit in the orders and concomitant classical elements, but this underscores the importance he placed upon them. Their use in ecclesiastical architecture had already been established, having made the transition from pagan and secular sources in ancient Rome, where they had been used for the glorification of the gods or of the state. Palladio's now fully Roman classical vocabulary was simply the only means of ornamentation acceptable to him. His genius lay in his inventiveness and creativity in assembling the elements of this vocabulary in order to shape space, articulate solid, and modulate light with elegance, grace, and beauty.

Fig. 9. Interior, San Giorgio Maggiore (photograph by the author)

The work that secured Palladio such a prominent place in the history of architecture is not made of brick or stone, however. It is his treatise, *I quattro libri dell'architettura*, first published in 1570 but translated and republished myriad times since, which ensured that his influence would be felt centuries after his death in 1580. It became the *de facto* primary source book for classical architecture, since Palladio included a plethora of painstakingly detailed and amply dimensioned architectural drawings, from designs to details, which would be the inspiration for many later architects.[8] He also presents his design philosophies in substantial detail, and the importance he placed on ratio and proportion (recall "the relationship of the whole to the parts, and of the parts among themselves and to the whole") cannot be overstated. The specifications of the five orders in book I are exhaustively dimensioned in terms of a module based on the column diameter, in some cases to such precision as to be beyond what could actually be achieved or perceived, which underscores his theoretical approach.[9] Later in book I Palladio lists his seven preferred room types, namely the circle, the square, and rectangles with the following length-to-width ratios: $\sqrt{2}$:1, 4:3, 3:2, 5:3, and 2:1.[10] Of course, Palladio's own designs in book II, as well as those of Roman buildings in book IV, are ready made for creative borrowing and extensive analysis.[11]

Palladio's later works in Vicenza include Palazzo Valmarana (begun c. 1565), his first palace with a colossal order; Palazzo Barbarano (begun c. 1570); Palazzo Porto Breganze (begun c. 1571, fig. 10); Loggia del Capitaniato (begun c. 1571, fig. 11), which is located across the piazza from the Basilica; and Teatro Olimpico (begun c. 1580). The last of these Palladio designed for the Accademia Olimpica, of which he was one of the earliest members, and he completed his design for the theater just prior to his death in August of 1580.

Fig. 10. Palazzo Porto Breganze
(photograph by the author)

Fig. 11. Loggia del Capitaniato
(photograph by the author)

Palladio and his wife, Allegradonna, had five children, three of whom survived their father. Palladio's body was buried in the Dominican church of Santa Corona in Vicenza, but his remains were exhumed in the nineteenth century and moved to prominent tomb in a new, neoclassical civic cemetery.

Palladio certainly benefited from the renaissance of intellectual thought that surrounded him in cinquecento Italy. The studies occurring in the various circles to which Palladio belonged over the years, such as those associated with Alvise Cornaro, Giangiorno Trissino, and Daniele Barbaro, were widely ranging. Palladio himself was at times stone mason, architect, engineer, archaeologist and architectural historian. Through meticulous research in Rome and elsewhere, he was able to develop an authentic classical vocabulary from ancient and contemporary sources, which he incorporated with seemingly boundless care and ingenuity in order to design an impressively large number of exceptionally beautiful and sturdy buildings. In the final analysis, however, the exceptional beauty of his architecture depends on an inborn artistic ability that cannot be quantified or otherwise explained by the influences of those around him. There is a huge difference between classically true and truly beautiful, and it is Palladio's innate mastery of aesthetics that is his greatest legacy.

Acknowledgment

This biography of Palladio was originally written for "The Year of Palladio" website of the Institute for Classical Architecture and Classical America (http://www.classicist.org/resources/year-of-palladio/), and the bibliography accompanying the original includes several sources for further reading. The endnotes and bibliography have now been expanded from the original version to highlight numerous publications concerning relationships between architecture and mathematics in Palladio's oeuvre.

Notes

1. The first two quotations are from [Puppi 1975: 27]; the third is from [Boucher 1994: 20].
2. This first-hand architectural information eventually became source material for Palladio's first two publications, *Le antichità di Roma* and *Descrizione della chiese di Roma*, both published in 1554, as well as for book IV of his treatise, *I quattro libri dell'architettura*, published in 1570.
3. Palladio's treatise, book I, chapter 1, p. 6–7 [Palladio 1997: 7]; similar statements are found in book II, chapters 1–2, p. 3–4 [Palladio 1997: 77–78]. For an example from Palladio's work, at Villa Cornaro not only do the exterior columns reflect the placement of the interior walls and columns; the main sala's length-to-width ratio equals not only its width-to-height ratio but also the ratio of the lower-storey column diameter to the upper-storey column diameter. See [Mitrović and Wassell 2006: 28–29, 46]; the ratios cited are based on dimensions from their 2003 survey of Villa Cornaro, on which their book is based.
4. Note Puppi's quotes of Barbaro in the following: "For Barbaro the scientific basis of knowledge was to be found in mathematics, … [and] he concludes that 'some arts have more of science and others less', and the 'more worthy' are 'those wherein the art of numeracy, geometry, and mathematics is required'"; see [Puppi 1975: 18]. Puppi cites his quotes as D. Barbaro, 1556, p. 7, i.e., from Barbaro's translation and commentary of Vitruvius [Barbaro 1556].
5. [Wittkower 1988: 104 ff]. Wittkower bases his analysis on the dimensions given in Palladio's treatise. Two articles providing more in-depth analyses related to Wittkower's appear in the *Journal of the Society of Architectural Historians*: [Howard and Longair 1982] and [Mitrović 1990]. A more exhaustive study of potential proportional sources for Palladio is available in [March 1998]; indeed, Palladio's corpus certainly supports March's claim: "In truth the Renaissance might be called the era of conspicuous erudition in which patrons, scholars and artists displayed their breadth of classical learning in various works and commissions" [March 1998: xii]. For example, although Palladio's use of *ad triangulum* references (inherently involving $\sqrt{3}$) on the Villa Rotonda cannot be explained through Wittkowerian means, the length-to-width ratio of the four large corner rooms (26:15) is a canonical ancient Greek approximation for $\sqrt{3}$, as Mitrović stated in his first *JSAH* article [Mitrović 1990: 285]. Moreover, the diameter of the central circular room, 30, completes a virtual 30°-60°-90° triangle. (The dimensions 15, 26, and 30 are those given by Palladio in his treatise, book II, chapter 3, p. 19 [Palladio 1997: 95].) To state this using modern notation, the angles of a triangle with side lengths 15, 26, and 30 are, to the hundredth of a degree, 30.00°, 60.07°, and 89.93°; cf. [Wassell 1999: 124–125].
6. See, for example, two articles in the *Journal of the Society of Architectural Historians*: [Millon 1972] and [Payne 1994].
7. See the chapter "Palladianism Today" in [Mitrović 2004: 171–187], where he carefully analyzes and counters arguments against the use of classicism today, especially regarding the issue of appropriateness to time, and offers several examples of contemporary Palladian architecture.
8. The principal competitor, over the years, to Palladio's treatise has been [Vignola 1563]. While some prefer Vignola's canon of the five orders, his treatise lacked the plethora of drawings available in books II through IV of *I quattro libri dell'architettura*.
9. See book I, chapters 12–19 of [Palladio 1997]; cf. [Mitrović 2004: 149].
10. See book I, chapter 21, p. 52 of [Palladio 1997: 57]; the full text is:

    There are seven types of room that are the most beautiful and well proportioned and turn out better: they can be made circular, though these are rare; or square; or their length will equal the diagonal of the square of the breadth; or a square and a third; or a square and a half; or a square and two-thirds; or two squares.

Palladio also gives constructions for the arithmetic, geometric, and harmonic means in book I, chapter 23, p. 53–54 [Palladio 1997: 58–59], and he recommends choosing from them to find a vaulted room's height given its length and width.

11. Palladio included his Basilica amongst the public architecture in book III, which is otherwise dominated by bridge design.

Bibliography

Sources published before 1800

ALBERTI, Leon Battista. 1486. *De re aedificatoria*. Florence. English trans. Joseph Rykwert, Neil Leach and Robert Tavernor, *On the Art of Building in Ten Books* (Cambridge, Mass.: MIT Press, 1988).

BARBARO, Daniele. 1556. *I dieci libri dell'architettura di M. Vitruvio, tradotti et commentati*. Venice.

BELLI, Silvio. 1573. *Della Proportione, et Proportionalità, Communi Passioni del Quanto, Libri Tre*. Venice. English trans. and commentary Stephen R. Wassell and Kim Williams, with a foreword by Lionel March, *On Ratio and Proportion, The Common Properties of Quantity* (Florence: Kim Williams Books, 2002).

BERTOTTI–SCAMOZZI, Ottavio. 1786. *Le fabbriche e i disegni di Andrea Palladio*. Venice.

PALLADIO, Andrea. 1570. *I quattro libri dell'architettura*. Venice. English trans. Robert Tavernor and Richard Schofield, *The Four Books on Architecture* (Cambridge, Mass.: MIT Press, 1997).

———. 1554. *Le antichità di Roma*. Rome; 1554 *Descrizione della chiese di Roma*. Rome. English trans. into a single volume by Vaughan Hart and Peter Hicks, *Palladio's Rome* (New Haven: Yale University Press, 2006).

SERLIO, Sebastiano. 1566. *Tutte l'opere d'architettura et prospetiva*. Venice. English trans. Vaughan Hart and Peter Hicks, *Sebastiano Serlio On Architecture* (New Haven: Yale University Press, 1996).

VIGNOLA, Giacopo Barozzi da. 1563. *Regola delli cinque ordini*. Rome. English trans. with a reprint of the second edition (Rome, 1572) Branko Mitrović, *Canon of the Five Orders of Architecture* (New York: Acanthus Press, 1999).

VITRUVIUS. *De architectura*. English trans.: Frank Granger, *On architecture*, (Cambridge, MA.: Harvard University Press, 1931); Morris Hickey-Morgan, *The Ten Books on Architecture* (New York: Dover, 1960); Ingrid D. Rowland and Thomas Noble Howe, *Ten Books on Architecture* (New York: Cambridge University Press, 1999).

Sources published after 1800

ACKERMAN, James. 1966. *Palladio*. Harmondsworth: Penguin.

BELTRAMINI, Guido, Antonio PADOAN, Howard BURNS and Pino GUIDOLOTTI. 2001. *Andrea Palladio: The Complete Illustrated Works*. New York: Universe Publishing.

BOUCHER, Bruce. 1994. *Andrea Palladio: The Architect in his Time*. New York: Abbeville Press.

BURNS, Howard, Lynda FAIRBAIRN and Bruce BOUCHER. 1975. *Andrea Palladio 1508–1580: The Portico and the Farmyard*. London: Arts Council of Great Britain.

COOPER, Tracy E. 2006. *Palladio's Venice: Architecture and Society in a Renaissance Republic*. New Haven: Yale University Press.

FLETCHER, Rachel. 2000. Golden Proportions in a Great House: Palladio's Villa Emo. Pp. 73-85 in *Nexus III: Architecture and Mathematics*. Kim Williams, ed. Pisa: Pacini Editore.

———. Palladio's Villa Emo: The Golden Proportion Hypothesis Defended. *Nexus Network Journal* 3, 2 (Summer-Autumn 2001): 105-12.

FOWLER, David. 1999. *The Mathematics of Plato's Academy: A New Reconstruction*. New York: Oxford University Press.

GABLE, Sally and Carl I. GABLE. 2005. *Palladian Days: Finding a New Life in a Venetian Country House*. New York: Knopf Publishing.

GIACONI, Giovanni and Kim WILLIAMS. 2003. *The Villas of Palladio*. New York: Princeton Architectural Press.

HERSEY, George and Richard FREEDMAN. 1992. *Possible Palladian Villas (Plus a Few Instructively Impossible Ones)*. Cambridge, Mass.: MIT Press.

HOWARD, Deborah and Malcolm LONGAIR. 1982. Harmonic Proportion and Palladio's Quattro Libri. *Journal of the Society of Architectural Historians* **41**, 2 (May 1982): 116–143.

LEWIS, Douglas. 1981. *The Drawings of Andrea Palladio*. Washington D.C.: International Exhibitions Foundation; second edition, 2000. New Orleans: Martin & St. Martin.

MARCH, Lionel. 1998. *Architectonics of Humanism: Essays on Number in Architecture*. London: Academy Editions.

———. 1999. Architectonics of proportion: a shape grammatical depiction of classical theory," *Environment and Planning 'B': Planning and Design* **26**: 91–100.

———. 1999. Architectonics of proportion: historical and mathematical grounds. *Environment and Planning 'B': Planning and Design* **26**: 447–454.

———. 2001. Palladio's Villa Emo: The Golden Proportion Hypothesis Rebutted. *Nexus Network Journal* **3**, 2 (Summer-Autumn 2001): 85-104.

MILLON, Henry. 1972. Rudolf Wittkower, Architectural Principles in the Age of Humanism, Its Influence on the Development and Interpretation of Modern Architecture. *Journal of the Society of Architectural Historians* **31**, 2 (May 1972): 83–91.

MITCHELL, William J. 1990. *The Logic of Architecture: Design, Computation and Cognition*. Cambridge, MA: MIT Press.

MITROVIĆ, Branko. 1990. Palladio's Theory of Proportions and the Second Book of the *Quattro Libri dell'Architettura*. *Journal of the Society of Architectural Historians* **49**, 3 (September 1990): 272–292.

———. 1999. Palladio's Theory of Classical Orders in the First Book of *I Quattro Libri Dell'Architettura*. *Architectural History* **42**: 110–140.

———. 2001. A Palladian Palinode: Reassessing Rudolf Wittkower's Architectural Principles in the Age of Humanism. *Architectura* **31**: 113–131.

———. 2004. *Learning from Palladio*. New York: Norton.

MITROVIĆ, Branko and Stephen R. WASSELL. 2006. *Andrea Palladio: Villa Cornaro in Piombino Dese*. New York: Acanthus Press.

PAYNE, Alina A. 1994. Rudolf Wittkower and Architectural Principles in the Age of Modernism. *Journal of the Society of Architectural Historians* **53**, 3 (September 1994): 322–342.

PUPPI, Lionello. 1973. *Andrea Palladio*. Milan: Electa. English trans. Pearl Sanders, *Andrea Palladio* (Boston: New York Graphic Society, 1975).

ROBISON, Elwin C. 1998–1999. Structural Implications in Palladio's Use of Harmonic Proportions. *Annali d'architettura* **10-11**: 175–182.

ROSE, P. L. 1975. *The Italian Renaissance of Mathematics: Studies on Humanists and Mathematicians from Petrach to Galileo*. Geneva: Librarie Droz.

ROWE, Colin. 1982. *The Mathematics of the Ideal Villa and Other Essays*. Cambridge, MA: MIT Press. ("The Mathematics of the Ideal Villa" was first published in *Architectural Review* **101** (1947): 101–104.)

RYBCZYNSKI, Witold. 2002. *The Perfect House: A Journey with Renaissance Master Andrea Palladio*. New York: Scribner.

RYKWERT, Joseph. 1999. *The Palladian Ideal*. New York: Rizzoli.

SPINADEL, Vera W. de. 1999. "Triangulature" in Andrea Palladio. *Nexus Network Journal* **1**: 117–120.

STINY, George and William J. Mitchell. 1978. The Palladian Grammar. *Environment and Planning 'B': Planning and Design* **5**: 5–18.

TAVERNOR, Robert. 1991. *Palladio and Palladianism*. London: Thames and Hudson.

WASSELL, Stephen R. 1998. The Mathematics of Palladio's Villas. Pp. 173-186 in *Nexus II: Architecture and Mathematics*. Kim Williams, ed. Fucecchio (Florence), Italy: Edizioni Dell'Erba.

——. 1999. Mathematics of Palladio's Villas: Workshop '98. *Nexus Network Journal* 1: 121–128.

WITTKOWER, Rudolph. 1988. *Architectural Principles in the Age of Humanism*. London/New York: Academy Editions/St. Martin's Press. (First published as vol. 19 of Studies of the Warburg Institute in 1949.)

About the author

Stephen R. Wassell received a B.S. in architecture in 1984, a Ph.D. in mathematics (mathematical physics) in 1990, and an M.C.S. in computer science in 1999, all from the University of Virginia. He is a Professor of Mathematical Sciences at Sweet Briar College, where he joined the faculty in 1990. Steve's primary research focus is on the relationships between architecture and mathematics. He has co-authored two books, one with Kim Williams entitled *On Ratio and Proportion* (a translation and commentary of Silvio Belli, *Della proportione et proportionalità*), and one with Branko Mitrović entitled *Andrea Palladio: Villa Cornaro in Piombino Dese* (see review in this issue). Steve's overall aim is to explore and extol the mathematics of beauty and the beauty of mathematics.

Lionel March

Martin Centre for Architectural
and Urban Studies
University of Cambridge
1-5 Scroope Terrace
Cambridge, CB2 1PX, UK
lmarch@ucla.edu

Keywords: Andrea Palladio,
arithmetic mean, geometric
mean, harmonic mean, number
symbolism, Vitruvius, Alberti,
triangular numbers, rational
approximations

Research

Palladio, Pythagoreanism and Renaissance Mathematics

Abstract. This paper examines the proportional qualities of Palazzo Antonini in terms of contemporary, Renaissance mathematics. It reveals that Palladio was either a masterful arithmetician, or a serendipitous genius.

Mathematical studies of Palladio's work are most often written from a modern view point. One serious anachronism is the use of the decimal notation. The problem here is that the decimals were not introduced until Simon Stevin published his arithmetic in 1585. Conversion of ratios into decimals obfuscates the rational number system employed in Palladio's time. Today, a rational number is represented by p/q where numerator and denominator are whole numbers.

Numbers were conceived with a rich variety of character unfamiliar to many commentators today. Modern writers are more aware of powers and roots, but less of polygonal and pyramidal numbers, less of numbers as lines, oblongs, beams, bricks and scalenes; more of arithmetic, geometric and harmonic means, but less of the seven to eight other classical means known to Renaissance arithmeticians; more of odd and even numbers, but less of the finer classifications current in Palladio's period; less of the tenfold description of rational numbers; and less of alphanumeric equivalences, encodings and cabalistic manipulations.

Suppose a contemporary of Palladio, (1508-1580) such as Girolamo Cardano (1501-1576), were to have examined the *Quattro libri dell'architettura* [Palladio 1570; Tavernor and Schofield 1997] from an arithmetical viewpoint. Cardano is cited in Barbaro's commentaries on Vitruvius [1556; 1567], and Cardano confirms this citation in *The Book of My Life* [Grafton 2002]. In this autobiography, Cardano writes that he was taught arithmetic by his father. From his own description it can be assumed that what he learned as a child would have been a blend of practical *abachista* computations and arcane, Pythagorean arithmetic – Nicomachus and/or Boethius. He mastered Euclid (Books I - VI) at the age of twelve. A polymath, he was a prolific author on mathematics, astrology, medicine, publishing his algebraic masterwork *Ars Magna* in 1570, the year also of the *Quattro libri*. This paper looks selectively at the *Quattro libri* through the lens of such a contemporary.

On the title page Book I sets out "rules [*avertimento*] essential to building" [Tavernor and Scholfield 1997: 1]. In chapter 21, Palladio numerates seven room shapes "that are most beautiful and well proportioned", of which, aside from the circle, six are rectangular: "the square; or their length will equal the diagonal of the square of the breadth; or a square and a third; or a square and a half; or a square and two thirds; or two squares" [Tavernor and Schofield 1997: 57]. A Renaissance arithmetician, particularly a player of *rithmomachia* like Cardano [March 1998: 49ff], would instantly identify these ratios with successive pairs of triangular numbers (fig. 1):

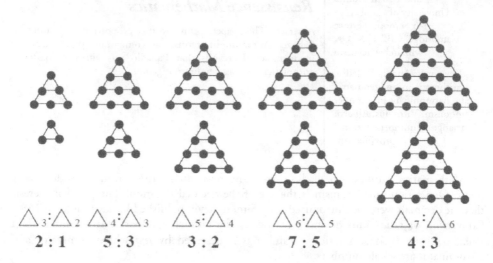

$$\triangle_3 : \triangle_2 \quad \triangle_4 : \triangle_3 \quad \triangle_5 : \triangle_4 \quad \triangle_6 : \triangle_5 \quad \triangle_7 : \triangle_6$$

2 : 1   5 : 3   3 : 2   7 : 5   4 : 3

Fig. 1. Ratios between successive pairs of triangular numbers:
6 : 3 :: 2 : 1; 10 : 6 :: 5 : 3; 15 : 10 :: 3 : 2; 21 : 15 :: 7 : 5 ; 28 : 21 :: 4 : 3

The ratio 7 : 5 is a classic proxy for $\sqrt{2}$: 1. Sir Thomas Heath notes that "Plato, and even Pythagoreans, were familiar with 7/5 as an approximation for $\sqrt{2}$" [1986, II: 119]. Alberti in *Ludi Matematici* comments, in reference to Pythagoras's theorem, that people who make a right-angled triangle from two sides of five feet and a hypotenuse of seven "are wrong, because their squares do not respond to the total; but ignore one fiftieth part" [Rinaldi 1980: 52]. By Pythagoras's theorem:

$$5^2 + 5^2 = 50 > 49 = 7^2.$$

There is no rational number which represents $\sqrt{2}$. It is said to be irrational, inexpressible. Inexpressible roots can be approximated using a well known relation between the classic means: arithmetic mean > geometric mean > harmonic mean, which when applied to rational extremes p > q, yields

$$(p + q)/2 > \sqrt{pq} > 2pq/(p + q).$$

Thus between $p = 2$ and $q = 1$

$$3/2 > \sqrt{2} > 4/3.$$

Now, a value between 3/2 and 4/3 was known to be (3 + 4)/(2 + 3) = 7/5, or the sum of the numerators over the sum of the denominators [Chuquet 1484; see Flegg et al. 1985].

In searching for a rational proxy for $\sqrt{2}$, it is seen that two other numbers 3/2 and 4/3 are implicated. Thus, three of Palladio's proportions are evoked with two more marked by extremes 2/1 and 1/1. Only the ratio 5 : 3 is not evoked. However, 5/3 is the contra-harmonic mean between 2 and 1, as is illustrated in Nicomachus's *Introduction to*

Arithmetic (II.28) as 6 : 5 : 3, or 2 : 5/3 : 1. This is the fourth mean defined among ten cited by Nicomachus [D'Ooge 1938]. Any competent Renaissance arithmetician would be familiar with this mean.

Turn now to Book II, which "contains drawings of many houses designed by him [Palladio]". In Chapter 1 Palladio declares that "in the previous book I explained everything that seemed most worthy of attention in the construction of public buildings and private houses so that the resulting work may be beautiful, graceful, and permanent ..." [Tavernor and Schofield 1997: 77]. The first scheme illustrated, Chapter 3, is Palazzo Antonini in Udine (fig. 2). It might be expected that this scheme would be an exemplar of the 'rules' set out in the first book. A cursory glance, however, shows that only the corner rooms on the garden side fit the canon. They are square, 17 : 17 :: 1 : 1, the ratio of equality.

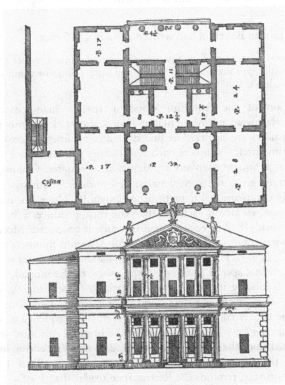

Fig. 2. Woodcut of the Palazzo Antonio from Palladio's *Quattro libri* [Tavernor and Schofield 1997: Bk. II, ch. III (p. 80)]

Looking at the wood cut showing plan and street elevation, a Renaissance arithmetician would take particular interest in the numbers. There are fourteen distinct numbers on the plate measuring dimensions in Vicentine feet. In no way do these numbers make it possible to compute the overall dimensions of the scheme. Notably absent are the thicknesses of the walls. It suggests that the numbers are independent of material, that is, they tend towards the Pythagorean 'conceptual and immaterial'. They relate to empty space, not solid matter. Someone familiar with Nicomachus might recall his words:

All that has by nature with systematic method been arranged in the universe seems both in part and as a whole to have been determined and ordered in accordance with number, by the forethought and the mind of him that created all things; for the pattern was fixed, like a preliminary sketch, by the domination of number pre-existent in the mind of the world-creating God, number conceptual only and immaterial in every way, but at the same time the true and the eternal essence, so that with reference to it, as to an artistic plan, should be created all [things] (Book I.6).

Nicomachus (II.19), with number in mind as source of creation, reports that:

... things are made up of warring and opposite elements and have in all likelihood taken on harmony – and harmony always arises from opposites; for harmony is the unification of the diverse and the reconciliation of the contrary-minded ....

In this spirit, Johann Reuchlin had written early in 1516 that:

... [the] world is more perfect, the more it contains many modes of numbering: equality, inequality; squares, cubes; length and area; primes or compound numbers [M & S Goodman, 1983].

An examination of the Antonini woodcut reveals many modes of numbering appreciated by a Renaissance arithmetician. Fig. 3 shows the numbers reduced to their factors. Prime numbers are depicted as one dimensional lines, since 1 is their sole factor aside from the prime itself. Numbers with two factors greater than 1 are depicted as two dimensional oblongs. Square numbers are depicted as squares. Oblongs with sides in the relation $p + 1 : p$ are given the special name *heteromecic*. With more than three factors greater than 1, three dimensional depictions are possible. Three distinct factors produce a 'scalene' solid. Two equal factors each less than the third produces a 'beam'; or each greater than the third, a 'brick'. If all three are equal, the cube is produced. More than three factors lead to multiple representations of the number. All Antonini numbers are generated by the following factors 2, 3, 5, 7, 11, 17, 19. In the Renaissance all primes were considered to be odd. The dyad, 2, held a special position as 'the other' to the monad, 1, 'the same'. It will be noted that 13 is missing in the sequence of primes from 3 to 19. In the woodcut the dimension 1 & 7 1/2 is inscribed on a second floor column. This dimension is in feet and inches and may be converted to 1 5/8, or 13/8 feet. The sequence of odd primes is thus complete. (The number 13 was not included in the original Antonini set, fig. 3, because it had to be computed from the one measure in fig. 2 to use Vincentine inches.) The span of odd numbers in the augmented Antonini set ranges from 3 to 19. Barbaro, in his commentary on Vitruvius, reminds the Renaissance reader that $3 + 5 = 8 = 2^3$; $7 + 9 + 11 = 27 = 3^3$; $13 + 15 + 17 + 19 = 64 = 4^3$, and that $1^3 + 2^3 + 3^3 + 4^3 = 100 = 10^2$, a relationship remarked upon by Alberti (IX.5). It may not be a coincidence that the pair of small central rooms in the Palazzo each has an area of 100 square feet.

Other modes of numbering are illustrated in fig. 4. For Palladio and his numerate contemporaries, numbers were 'perfect' or not. In the Vitruvian tradition, 6 and 28 were known to be perfect since $6 = 1·2·3$ and $1 + 2 + 3 = 6$; while $28 = 1·2·2·7$ and $1 + 2 + 4 + 7 + 14 = 28$. Numbers which are not perfect are classified as 'abundant' or 'deficient' depending on whether the sum of factors exceeds or falls short of the number.

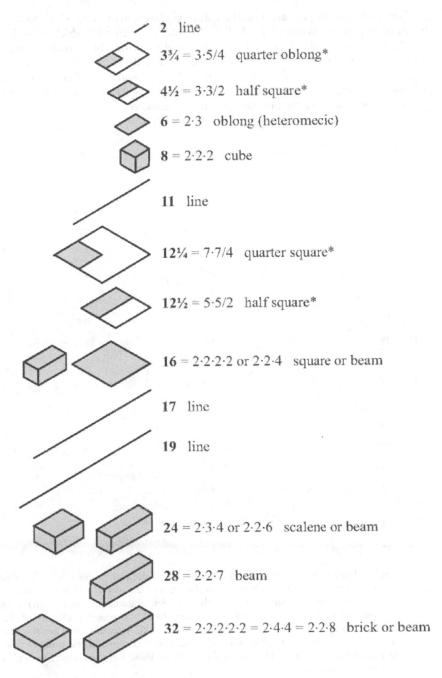

2 line

3¾ = 3·5/4   quarter oblong*

4½ = 3·3/2   half square*

6 = 2·3   oblong (heteromecic)

8 = 2·2·2   cube

11 line

12¼ = 7·7/4   quarter square*

12½ = 5·5/2   half square*

16 = 2·2·2·2 or 2·2·4   square or beam

17 line

19 line

24 = 2·3·4 or 2·2·6   scalene or beam

28 = 2·2·7   beam

32 = 2·2·2·2·2 = 2·4·4 = 2·2·8   brick or beam

Fig. 3. The Antonini number set as conceived by Renaissance arithmeticians depicting, in isometric projection, factors as lines, oblongs and solids (Nicomachus II.15-17). Numbers marked * are presented as improper fractions

Even numbers were divided into three distinct classes: even-even, which today would be described as positive powers of 2 (2, 4, 8, 16, 32, ...); even-odd, which when halved give an odd number (6, 10, 14, 18, 22, ...); and odd-even, in which halving leaves an even number (12, 20, 24, 28, 36, ...). Odd numbers were either prime (3, 5, 7, 11, 13, 17, 19, ...), or composite (9, 15, 21, 25, 27, 35, ...). Then again, numbers were visualized as polygonal. Triangular numbers are illustrated above in fig. 1. Square numbers may be seen in the same way, as can cube numbers in three dimensions.

numerator					
2				prime*	
3¾→15	1+3+5 = 9	deficient	3·5	odd-composite	△ 5
4½→ 9	1+3 = 4	deficient	3·3	odd-composite	◇ 3
6	1+2+3 = 6	perfect	2·3	even-odd	△ 3
8	1+2+4 = 7	deficient	2·2·2	even-even	
11				prime	
12¼→49	1+7 = 8	deficient	7·7	odd-composite	◇ 7
12½→25	1+5 = 6	deficient	5·5	odd-composite	◇ 5
16	1+2+4+8 = 15	deficient	2·2·2·2	even-even	◇ 4
17				prime	
19				prime	
24	1+2+3+4+6+8+12 = 36	abundant	2·2·2·3	odd-even	
28	1+2+4+7+14 = 28	perfect	2·2·7	odd-even	△ 7
32	1+2+4+8+16 = 31	deficient	2·2·2·2·2	even-even	

Fig. 4. Further Renaissance classification of the Antonini number set. The first column after the numbers in bold shows the sums of divisors by which numbers are said to be deficient – sum less than the number, abundant – sum greater than number, or perfect – sum equal to number. Note the presence of two perfect numbers, 6 and 28. The next perfect number is 496. Vitruvius cites these. There then follows a column of factors and a classification into even and odd number types (Nicomachus II.8 - 13). The final column shows those numbers which are either triangular or square.

* The even number 2 had a special place as the dyad and was rarely ranked with the other primes all of which are odd

Once the Renaissance classifications are appreciated, the Antonini set is seen to satisfy the Nicomachean requirement, in harmonious 'world-making', for contrariness and diversity among its members; and Reuchlin's Pythagorean ingredient for perfection in its many modes of numbering. Pre-existent numbering ensures both harmony and perfection in creative activity.

	2	3¾	4½	6	8	11	12¼	12½	16	17	19	24	28	32
I		3¾			8		12¼							
I				6		11			16					
I				6						17			28	
I				6							19			32
I									16			24		32
I												24	28	32
II	2				8									32
II			4½	6	8									
II					8				16					32
X	2			6	8									
X	2									17	19			
X		3¾					12¼		16					
X			4½					12½		17				
X				6		11				17				
X					8	11					19			
X					8							24		32
X						11				17			28	

Table 1. Proportionate relations between numbers span the Antonini set. Three Nicomachean means are noted: I, arithmetic; II, geometric; and X, the tenth mean in which $b = a - c \ (a > b > c)$

A study of the Antonini set also shows that the numbers are related proportionately in terms of the Nicomachean means (Table 1). Every number is related to at least two other and in some case several pairs. The relationships 'span' the set from 2 to 32. The simple Nicomachean tenth mean, X, is easily found in the set although it is not usually identified as such in modern studies. It does appear in Fibonacci references since its terms $a > b > c$ satisfy $a = b + c$, the condition between three consecutive terms of a Fibonacci sequence.

However, in determining convergents to the extreme and mean ratio, this relationship was not recorded explicitly until Kepler in the seventeenth century [Herz-Fischler 1998].

Returning to the plan, the spatial arrangement of the numbers in the wood cut may be examined (fig. 5). The ratios are oriented (left to right | front to back). They are classified in the contemporary manner [Belli 1573; Wassell and Williams 2003] showing yet more diversity and contrasts among the numbers of the Antonini set. As has been observed, only the square corner rooms (1 : 1) precisely match the seven canonic proportions. The garden loggia is assumed to be 32 : 17, but Palladio avoids dimensioning it. In any event, it is technically not a room [sala]. There are two rooms at the centre of the sides of the building (24 : 17). Earlier it was demonstrated that

$$3/2 > \sqrt{2} > 4/3.$$

Taking 3/2 and 4/3 as new extremes, another iteration of this same procedure gives improved convergents $17/12 > \sqrt{2} > 24/17$.

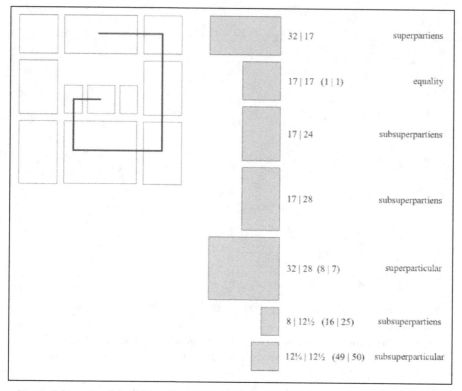

32 | 17	superpartiens
17 | 17 (1 | 1)	equality
17 | 24	subsuperpartiens
17 | 28	subsuperpartiens
32 | 28 (8 | 7)	superparticular
8 | 12½ (16 | 25)	subsuperpartiens
12¼ | 12½ (49 | 50)	subsuperparticular

Fig. 5. Schematic of the Palazzo Antonini plan showing room dimensions in Vicentine feet. The ratios are oriented (left to right | front to back) and described in the contemporary manner. The first 'room' is the garden loggia and its dimensions are extrapolated from the principal hall (32 | 28) on the street side and the two side rooms, left and right, (17 | 17)

Thus these two side rooms conform, as rational proxy 24 : 17, to the canonic $\sqrt{2}$: 1. On the street, the two corner rooms are 28 : 17. The dimension of 28 evokes perfection, but its relation to the width 17 is not obvious. If Palladio had intended a room of proportion 5 : 3 of width 17, he would have computed its length as 17·5/3 = $28^{1}/_{3}$. Thus, for a room of width 17, a 5 : 3 room has a length of 28 to the nearest foot. This can be viewed as a sensible, practical adjustment.

The main hall is 32 : 28 :: 8 : 7. Like 7/5 for $\sqrt{2}$, 7/4 had traditionally been regarded as a rational proxy for $\sqrt{3}$. This implies a right angled triangle of sides 7 and 4. The sum of the squares is 49 + 16 = 65. Now $\sqrt{65}$ – the hypotenuse – is close to $\sqrt{64}$ = 8 in the same way that $\sqrt{50}$ is close to $\sqrt{49}$ (as in the analysis of 7 : 5 above). The ratio 8 : 7 is then seen to be the ratio of the side of an equilateral triangle to its altitude. But this ratio, geometrically, is precisely $\sqrt{4}$: $\sqrt{3}$. Thus since 8 : 7 is a rational proxy for $\sqrt{4}$: $\sqrt{3}$, the canonic ratio 4 : 3 appears in the guise of its roots, *radices*.

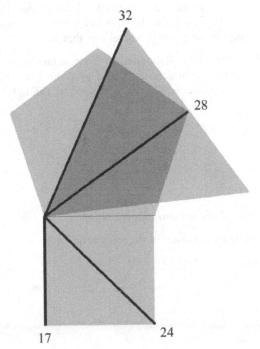

Fig. 6. Arithmo-geometric relations between the main room dimensions. Within tolerance, a square of sides 17 has a diagonal of 24, a regular pentagon with side 17 has a chord of length 28, and an equilateral triangle of altitude 28 has sides of length 32

There is a pair of small rooms whose proportion is 25 : 16 :: 5^2 : 4^2. If these were intended to be 4 : 3 rooms, then with a width of 8, their length would be 8·3/2 = 12. Yet Palladio adds a half foot in length. More likely, the architect evokes Alberti's *innatae correspondentae* (IX.6), those natural correspondences relating to geometrical figures. In this case, squares on the sides of the iconic 3, 4, 5 Pythagorean right angled triangle come to the *potentiae* 9, 16, 25. Note that the intercolumniation of the hexastyle portico is 9/2, which with 16/2 = 8, and 25/2, the sides of the central room, completes this Pythagorean triple at half scale. That Palladio does not accept the dimension 12 for a room of width 8 shows that his practice deliberately ranges beyond the canon.

The central, inner room has the ratio 50 : 49. As Alberti had observed this ratio is one-fiftieth short of the equality. Why does Palladio not make this central room a precise square? Perhaps Palladio hints at his interest in computations such as those related to roots, and √2 in particular. This central room can thus be seen as an 'ancestor', by virtue of computation, to the two central side rooms.

In *Architectonics of Humanism* [March 1998: 267] I draw attention to relationships between the areas of rooms in the case of Villa Malcontenta. Ratios between room areas in Palazzo Antonini reveal further evidence of Palladio's computational skills. Between the garden side corner rooms and the central rooms at each side, the ratio of areas is 24 : 17, or √2 : 1. The ratio between the areas of these central side rooms and the street side corner rooms is 28 : 24, or 7 : 6. The ratio of areas between these side rooms and the central hall is 32 : 17 which would appear to be the proportion of the garden loggia. Hidden in this dimension is the ratio 17 : 16, familiar at the time as a semitone used in tuning a lute. Two 17 : 16 rectangles long-side by long-side produces one that is 32 : 17.

Barbaro, with whom Palladio collaborated in his commentaries on Vitruvius, derives ratios through 'addition' and 'subtraction'. These procedures are equivalent to modern multiplication and division. Perhaps 32 : 17 is arrived at through such procedures. For example, the 'addition'

$$24 : 17 \text{ or } \sqrt{2} : 1$$
$$| \quad |$$
$$\underline{32 : 24} \text{ or } 4 \ : \ 3$$
$$32 : 17$$

clearly shows a relationship between two canonic ratios.

The ratio 28 : 17 bears examination in the same way:

$$24 : 17 \text{ or } \sqrt{2} : 1$$
$$| \quad |$$
$$\underline{28 : 24} \text{ or } 7 \ : \ 6$$
$$28 : 17$$

Now 7 : 6 is derived from the procedure using the inequality between the three classic means in determining a rational convergent for √4 : √3:

$$(1 + {}^4/_3) / 2 > \sqrt{4} : \sqrt{3} > 2.1.4/3 / (1 + {}^4/_3), \text{ or } 7/6 > \sqrt{4} : \sqrt{3} > 8/7.$$

From this, the ratio 28 : 17 is found to be the 'addition' of √2 : 1 and the root guise of 4 : 3, namely √4 : √3.

The ratio between the areas of the small central rooms is 12 $^1/_4$: 8 :: 49 : 32, or (32 + 17) : 32. Unlikely as it may have seemed there is this simple family bond between the small rooms and the proportion of the loggia. However, there is more: the 'subtraction' of a convergent to the cube root of 2, 9/7 (for this rational convergent see below) and the square root of 3, 7/4, gives

$$9 : 7 \text{ or} \qquad \sqrt[3]{2} : 1$$

$$\times$$

$$\underline{7 : 4} \text{ or} \qquad \underline{\sqrt{3} : 1}$$
$$49 : 36 \qquad \sqrt{3} : \sqrt[3]{2}$$

$$| \quad | \qquad\qquad | \quad |$$

$$\underline{36 : 32} \text{ or } 9 : 8, \quad \underline{3^2 : 2^3}$$

$$49 : 32 \qquad 3^2 \cdot \sqrt{3} : 2^3 \cdot \sqrt[3]{2}$$

This remarkable symmetrical relation, a rational proxy to 49 : 32, is the product of the *square* of 3 *and its square root* to the product of *the cube of 2 and its cube root*. In the fractional exponent notation introduced in mid-fourteenth century by Archbishop Bradwardine in *Tractatus proportionibus* [Crosby 1955] and Bishop Oresme in *De proportionibus proportionem* [Grant 1966]:

$$3^2 \cdot 3^{1/2} = 2^3 \cdot 2^{1/3}.$$

The skeptic will argue that $\sqrt[3]{2}$ is nowhere to be found in the Antonini set. However a more detailed examination of the tetrastyle hall reveals signs of the Delian cube – the problem of doubling the content of a cube (fig. 7).

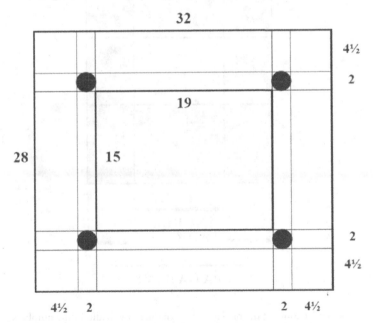

Fig. 7. The arrangement in the tetrastyle hall, Palazzo Antonini

The space between the four columns is found to be 19 by 15 feet. The ratio 19 : 15 is a good rational proxy for $\sqrt[3]{2}$: 1. The sequence of convergents starts with 4 : 3, which is too large, and the slightly too small 5 : 4, 'petrified' in Vitruvius' *pentadoron* and *tetradoron*, two stone blocks the first of which is practically double the volume of the other. Using the Chuquet method [Fowler 1987], (5 + 4) : (4 +3) = 9 : 7 is a better value (used in the computation above). This convergent is too large. A better value lies between 5 : 4 and 9: 7, say (9 + 5) : (7 + 4) = 14 : 11, though this is still a trifle too large. The next convergent is (14 + 5) : (11 + 4) = 19 : 15. The cube root of 2 is firmly placed in the most prominent location of the palazzo. It is even found in the internal elevation given that the columns are 19 feet tall (from the elevation) and 15 feet apart in the short direction. In this same direction the distance to the outside of the columns is 19 feet to the distance between of 15 feet. The cube root of two pervades the great hall.

The number 17 evokes the cabalistic tetragrammaton. The Hebrew name of god has the four letters HE VAV HE YOD. These letters have numerical equivalences which were popularized by Agrippa von Nettersheim in *De occulta philosophia*, 1533. HE = 5, VAV = 6, and YOD = 1 so that the four letter name sums to 5 + 6 + 5 + 1 = 17. Scholem (1974) describes this method of cabalistic gematria as the "small number" method using only the numbers 1 to 9. The plan of the palazzo is girdled on three sides by the divine number 17. The fourth side is marked by the perfect number 28. These boundary zones are patently auspicious (fig. 8).

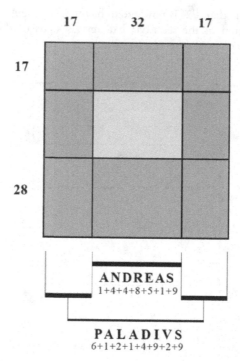

Fig. 8. The border zones of the plan and the signature of Palladio in the numbers 32 and 17 + 17 = 34

Palladio's name was contrived during his time with the humanist Gian Giorgio Trissino, "one of the most illustrious men of our time" [Tavernor and Schofield 1997: 5]. It is an artificial construction. In Latin the name was spelled PALADIVS, or PALLADIVS. It appears on the chapel at Maser with one L, and several documents are signed in this manner. Agrippa provides a nine-chamber arrangement of the 23 letter Latin alphabet (fig. 9). Palladio's Latin names sum to 32 and 34 in the small number method using the digits 1 - 9 only. That is, a number like 60 for P is reduced to 6, or 200 for V to 2. The double occurrence of the ratio 17 : 16 in the garden loggia may also reflect Palladio's encoded name. The most reduced contemporary encoding of Palladio's name is 3 + 4 = 7, 3 + 2 = 5. At the very heart of the palazzo is the 50 : 49 room with its familiar echoes of the historic 7 : 5 convergent to √2 : 1.

The augmented Antonini set includes the first seven odd primes (fig. 10). Framed by 19 and 3, the remaining five primes are 17, 13, 11, 7, 5. Is it a mere coincidence that 17.13 = 221 is the "large" number of ANDREAS, and 11.7.5 = 385 the number of PALADIVS? Remember, these reflect contemporary past-times. Palladio's name was deliberately, intentionally constructed, not given at birth. Barbaro's acknowledgment that Palladio was the Vitruvius of his age is arcanely confirmed in the encoding for VITRVVIVS = 1088 = 32·34 = ANDREAS PALADIVS.

A=1	B=2	C=3
K=10	L=20	M=30
T=100	V=200	X=300
D=4	E=5	F=6
N=40	O=50	P=60
Y=400	Z=500	
G=7	H=8	I=9
Q=70	R=80	S=90

Figure 9. The nine chamber numerical encoding of the Latin alphabet (Agrippa von Nettershaim, 1535)

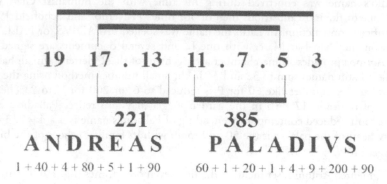

221

ANDREAS

$1 + 40 + 4 + 80 + 5 + 1 + 90$

385

PALADIVS

$60 + 1 + 20 + 1 + 4 + 9 + 200 + 90$

Fig. 10. The encoding of Palladio's Latin name using the full numbers from
Agrippa's nine squares

Perhaps, however, there is a simple, practical reason for all of this. In a meticulous and exhaustive study, Mitrović (2004) examines Palladio's buildings and the *Quattro libri* in the light of previous scholarship. Looking through other palazzo and villa plans, it is certain that powers and rational convergents to roots are discernable throughout. Towards the end of Book II there is a project for Giovanni Battista Garzadori which seems not to conform to the Book I canon or to Alberti's *innatae correspondentiae*. Principal rooms are proportioned 37 : 24, 37 : 32, 37 : 40, and 59 : 40. These dimensions are puzzling until the 37-35-12 Pythagorean triangle is recalled and then all becomes clear. Evident are the dimensions 59 = 12 + 35 + 12 feet for twice the width of the main hall and 12 + 20 + 16 = 48 feet, four times the short side of the triangle, for the combined length of the side rooms.

Returning to the Palazzo Antonini, the dimensions suggest the 17, 15, 8 triangle for the layout using a marked rope of length 17 + 15 + 8 = 40 feet. Fig. 11 shows the positions of the 'knots' in the rope if all six possible arrangements of the three side lengths are marked. These 'knots' divide the rope 8 + 7 + 2 + 6 + 2 + 7 + 8 = 40. Note that the triple 2 + 6 + 2 reflects the central dimensions of the hexastyle portico: column + intercolumniation + column. The rope is staked out to form a right angle taking the sides of the 17, 15, 8 triangle in any order by choosing two appropriate 'knots'.

Fig. 11. The six possible arrangements of the 17, 15, 8 Pythagorean triangle to mark
a 17 + 15 + 8 = 40 long rope for surveying purposes

A reader would have noticed that the woodcut shows the garden-side corner rooms as being rectangular whereas the numbers indicate square rooms, 17 by 17. If this is a cutter's error, the most likely number to replace 17 is 12 to make a 17 : 12 room (convergent to $\sqrt{2}$: 1). (Bertotti-Scamozzi's later survey suggests that this is correct). An application of the 17, 15, 8 triangle to laying out the palazzo plan reinforces this supposition (fig. 12). Having set up the triangle in the lower left corner to form a right angle, eight repeated applications of the triangle measure the length of the side wall and parallel cross wall (8·8 = 64 feet). In the upper left corner of fig. 12, it is shown that the hypotenuse, 17, can be swung round to measure the location of the cross wall (as well as the outermost columns of the façades). The next cross wall is 17 + 15 = 32 feet away. The procedure of eight applications is repeated to layout the new cross wall and parallel side wall. The front-to-back dimensions of the side rooms, 28, 24, and 12 feet, are clearly marked by the triangulations. The two internal walls parallel to the façade are thus located. The loggia now has the proportion 32 : 12 :: 8 : 3, a ratio, double 4 : 3, to be found in rooms at Maser.

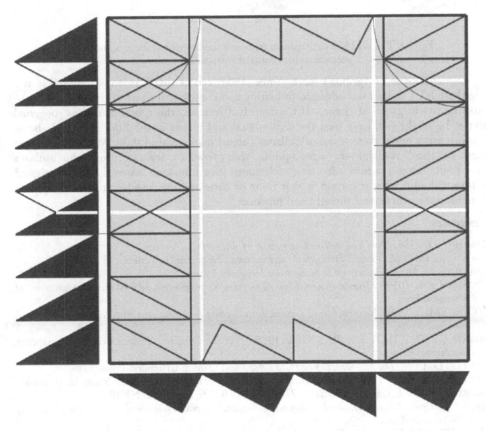

Fig. 12. Laying out the plan of the Palazzo Antonini using a 17, 15, 8 Pythagorean triangle

Next, the three inner rooms are to be marked out. The width of the two small side rooms is 8 feet, the small side of the 17, 15, 8 triangle (fig. 13). The length of $12^{1}/_{2}$ feet is easily measured by halving 17 + 8 = 25 feet. Likewise, the width of the central room is measured by dividing 17 + 15 + 17 = 49 feet in quarter, $12^1/_4$ feet. Finally, from the rope, half of 7 + 2 = 9 feet defines the side intercolumniations of the hexastyle portico. All plan dimensions are accounted for using this 17, 15, 8 Pythagorean triangle.

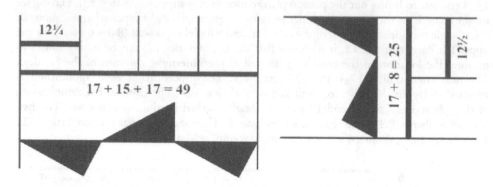

Fig. 13. The 17, 15, 8 Pythagorean triangle is used to set out dimensions of the smaller inner rooms of Palazzo Antonini

Cardano knew of Agrippa's work. Yates [1964] relates that Cardano "despised it as a trivial affair". Cardano was addicted to betting and contributed some of the first studies of probabilities in games of chance. If Cardano had reviewed the *Quattro libri*, as suggested above, he might well have seen the arithmetical and arcane possibilities indicated above. But just what might these reveal of Palladio's actual mathematical skills? What chance that parts of these readings are serendipitous interpretations separate from the author's intentions? That practical surveying techniques give mundane answers to the enigma? Without doubt, what is certain is that none of these readings are beyond contemporary Renaissance mathematical thought and practices.

References

BARBARO, D. 1556. *Dieci libri dell'archittetura di M. Vitruvio ... .* Venice.

————— . 1567. M. *Vitruvii Pollionis de architectura libri decem ... .* Venice.

CHUQUET, N. 1484. *Triparty en la Science des Nombres.* Lyon.

CROSBY, L. H. 1955. *Thomas Bradwardine. Tractatus de proportionibus.* Madison: University of Wsconsin Press.

D'OOGE, M. L. 1938. *Nicomachus of Gerasa. Introduction to Arithmetic.* Ann Arbor. University of Michigan Press.

FLEGG, G., C . HAY, and B. MOSS. 1985. *Nicolas Chuquet. Renaissance Mathematician.* Dordrecht: Reidel.

FOWLER, D. 1999. *The Mathematics of Plato's Academy.* Oxford: Clarendon Press.

GOODMAN, M and S. 1983. *Johann Reuchlin. On the Art of the Kabbalah.* New York: Abaris Books.

GRAFTON, A. 2002. *Girolamo Cardano. The Book of My Life.* New York: NYRB.

GRANT, E. 1966. *Nicole Oresme. De proportionibus and Ad pauca respicientes.* Madison: University of Wisconsin Press.

HEATH, T. 1986. Euclid. *The Thirteen Books of the Elements.* New York: Dover.

HERZ-FISCHLER, R. 1998. *A Mathematical History of the Golden Number.* New York: Dover.

MARCH, L. 1998. *Architectonics of Humanism.* Chichester, West Sussex: Academy Editions.

MITROVIĆ, B. 2004. *Learning from Palladio.* New York. W. W. Norton and Company.

PALLADIO, A. 1570. *I quattro libri dell'architettura.* Venice.

RINALDI, R. 1980. *Leon Battista Alberti. Ludi matematici.* Milan: Guanda.

SCHOLEM, G. 1974. *Kabbalah.* New York: Dorset Press.

TAVERNOR, R. and R. SCHOFIELD. 1997. *Andrea Palladio. The Four Books on Architecture.* Cambridge MA: The MIT Press.

WASSELL, S. R. and K. WILLIAMS. 2003. *Silvio Belli. On Ratio and Proportion.* Florence: Kim Williams Books.

YATES, F. 1964. *Giordano Bruno and the Hermetic Tradition.* London: Routledge & Kegan Paul.

About the author

Lionel March, Visiting Scholar, Martin Centre for Architectural and Urban Studies, University of Cambridge. Emeritus Professor of Design and Computation, University of California, Los Angeles. Founding editor, *Environment and Planning B*. General editor with Leslie Martin, *Cambridge Architectural and Urban Studies*. Co-author with Philip Steadman, *The Geometry of Environment*. Author, *Architectonics of Humanism, Essays on Number in Architecture*.

Buthayna H. Eilouti

Department of Architectural
Engineering
Jordan University of Science
and Technology
POB 3030
Irbid 22110, JORDAN
buthayna@umich.edu

Keywords: architectural
language, string recognition,
visual study, design
computation, Palladian
language, finite state automata,
FSA application, façade
morphology

Research

A Formal Language for Palladian Palazzo Façades Represented by a String Recognition Device

Abstract. This article represents an effort to reveal a new interpretation of the expression 'the architectural language of Palladian designs' that is closer to real linguistic paradigms than it usually means. Palladian designs exhibit a highly ordered and well articulated formal language comprised of a finite set of vocabulary elements in plan, elevation and volumetric treatment, together with an implicit set of mathematical rules for the arrangement of these rudimentary elements. The scope of this article is limited to the morphology of the façades of the first nine buildings shown in the second book of Palladio's treatise, specifically the palazzo designs that he presents in chapter three. The morphology is described in terms of a symbolic encoding system that is represented textually and graphically as a finite state automaton, the concept of which is borrowed from theories of formal languages and computation. The system helps to emphasize commonalities in façade languages and to propose a prototype for generating Palladian palazzo façade designs. The automaton-based encoding system may be developed to function as a base for a computerized façade encoder and decoder.

Regular language processing and recognition

Studies of formal languages, and their structural and computational representations in linguistics, computer science and related fields as well as their applications, are well-established (e.g., [Chomsky 1964]; [Aho and Ullman 1972]; [Salomaa 1973]; [Harrison 1978]; [Hopcroft and Ullman 1979]; [Linz 1997]; [Revesz 1983]). Furthermore, there are some research efforts that aim to expand their applicability to explain architectural design morphology (e.g., [Eilouti and Vakalo 1999]; [Eilouti 2005]; [Eilouti, 2007]). However, the applicability of formal language manipulation techniques to explain architectural language composition and processing is still inadequately explored. New layers of these techniques can be further investigated and employed to classify stylistic families of architectural artifacts, as seen in this article, which follows in the tradition of previous efforts to analyze Palladian designs based on methods related to computer science theory and practice (e.g., [Stiny and Mitchell 1978]; [Mitchell 1990]; [Hersey and Freeman 1992]).

In the context of formal languages, regular languages epitomize one of the simplest forms of language definition and processing. They consist of strings (words) constructed from a given alphabet of symbols (letters). A specific regular language consists of a set of accepted strings, where any given string is either accepted or rejected according to stipulated criteria that determine language membership. In the literature on computational application of regular languages, some simple devices are defined to help classify strings into languages. One of these devices is known as a finite state automaton (FSA).

A finite state automaton is an extremely restricted and simplified model of a computer that functions as a language recognition device. An FSA consists of a finite number of states, and specified transitions between states, with one state typically designated the initial state. The FSA is initialized with an input string, which it reads one symbol at a time. The automaton is in the initial state when it reads the first symbol, and then it enters a new state that is dependent on the current state and the symbol just read. The process of reading an input symbol and transitioning from state to state is repeated. This transition process can be both iterative and recursive, as states can be repeated as many times as required and can be called internally from any state to re-enter that state or enter another. When the symbols of the input string are exhausted, the reading process stops at one of the states. These terminal states can be either final states or not. Strings that end at final states are considered acceptable as members of the given language, and in this case, the automaton indicates its approval of the string it has read. Otherwise, if the string stops at one of the non-final states, the string is rejected and the automaton declares its disapproval of the string's membership to the language at hand. The language accepted by an automaton is the set of all strings it accepts. Any accepted string is said to be a member of the language that is defined by the automaton and any rejected string is labeled as not a member of that language [Lewis and Papadimitriou 1981].

One of the powers of FSAs is that they help generate enormous unforeseen strings. Another powerful aspect of FSAs lies in the capability to take the union, intersect and complement of languages in order to derive new languages. FSAs can manipulate symbols in order to enable users to model problems, test solutions and transform sets of symbol combinations into algorithms for which step-by-step procedures can be described. They have applications in computer algorithm building and processing as well as in the lexical analysis phase of a compiler.

The Mathematical and Graphical Representation of String Processing

Finite state automata of regular languages are generally classified into two types, the deterministic and the nondeterministic [Hopcroft and Ullman 1979]. While the transition from a state to another in the deterministic automaton is completely *determined* by the input symbol, the operation on some of the input symbols in the nondeterministic is probabilistic. In other words, while on any given input symbol a deterministic automaton must proceed from the current state to one and only one designated state, a nondeterministic may move to more than one state.

The essential mathematical representation of a deterministic FSA of a regular language recognition process is formulated as follows [Aho, Sethi and Ullman 1986]; [Lewis and Papadimitriou 1981]:

A FSA is a quintuple $M = (K, \Sigma, \delta, S, F)$, where

K: is a finite set of states (including final and non-final states)
Σ: is an alphabet of symbols
$S \in K$: is the initial state at which the recognition process starts
F: is the set of final states (subset of K) that determine acceptable strings
δ: is a transition function (from $K \times \Sigma$ to K) that maps a current state and a symbol to a new state.

A rule according to which the automaton M moves from a state (q ∈ K) to another state (q' ∈ K) is encoded into the transition function δ in terms of the input symbol and two states. For example, if the current state is q, the input symbol is σ, and the new state is q', then δ (q, σ) = q'.

FSAs are represented graphically as attributed state diagrams. The pictorial representation of the finite state diagram is a directed graph, where descriptive text is incorporated into the graphical components. States are represented by nodes, where a circle signifies a state of string processing, and double concentric circles signify final states which are also known as states of string acceptance. Arrows which connect two state circles are used to show the direction of string processing. The initial state where the process of string recognition begins is usually denoted by the symbol > (see fig. 1).

To demonstrate how FSAs function, let us consider the following example. Take the alphabet to consist of two symbols, X and Y. Let {X,Y}* denote all possible combinations of the rudimentary alphabet elements (i.e., all possible words made up of X's and Y's). Suppose we want to produce a finite state automaton M that generates the language L(M) consisting of the set of all strings in {X,Y}* that have an even number of Y's (including zero) and optionally any number of X's before, after or in between the Y's.

The finite state automaton M is illustrated by fig. 1. According to this FSA, at the start state q0, the input symbol 'X' can be added as many times as required (including zero times). Then, upon the addition of the symbol 'Y' the automaton moves to the state q1; as at state q0, zero or more 'X' symbols can be added. Next, the automaton can go back to the state q0 if another 'Y' symbol is added. The FSA transitions from q0 to q1, or vice versa, only upon the addition of a 'Y' symbol. When the string has been fully read, symbol by symbol, the process terminates. If the string contains an even number of Y's, the FSA will terminate in state q0, which is a final state in this example (denoted by the double circle), so that the string is accepted. If the string contains an odd number of Y's, the process terminates in q1, a non-final state, and so the string is not accepted.

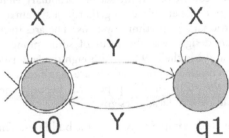

Fig. 1. A finite state automaton of the language consisting of the strings of {X,Y}* that include an even number of Y's

Examples of accepted strings, which thus belong to L(M), are: X, XXYY, XXXYXXY, and XYYYYX. They all have an even number of Y's. Examples of rejected strings are: XXYYY, XXXYXXYY, and XYYYX. These strings have an odd number of Y's and do not belong to L(M). This language can be represented as L(M) = (X*YX*Y)*. As above, an asterisk (*) means to repeat the symbol directly to the left (or group of symbols within parentheses) as many times as desired (≥ 0 times).

The four input/output values for the FSA's transition function $\delta(q,\sigma)$ are listed in the four rows of Table 1 (the first two columns are the input and the last is the output). Note that each input/output value of the transition function corresponds to an arrow from state to state (or from a state to itself) in the FSA.

Q	σ	$\delta(q, \sigma)$
q0	X	q0
q0	Y	q1
q1	X	q1
q1	Y	q0

Table 1. The table of input/output values for the transition function for the automaton of fig. 1

In the following sections, the concept and representations of the finite state automaton will be applied to a selected set of Palladian façades to help explore new layers of their underlying compositional and procedural languages.

The Palladian Design Study Sample

For the purposes of this study, the scope of analysis is limited to the designs illustrated in chapter three of the second book of Palladio's 'Four Books of Architecture' [Palladio 1997]. This chapter includes palazzos (houses in the city or town houses) that are designed by Andrea Palladio. It is noteworthy to point out that one of the designs illustrated in this chapter represents a villa rather than a palazzo. About the inclusion of this villa, which is the house of Monsignor Paolo Almerico (Villa Rotonda) in Vicenza, with the city houses explains Palladio "because it is so close to the city that one could say it is in the city itself" [Palladio 1997: 94].

All of the Palladian houses, exemplified by the ones selected in this study, exhibit a high level of order and consistency in their geometric language. A visual study of the palazzo designs reveals that they consist of a finite set of vocabulary elements in plans, elevations and massing. The components are placed together in a systematic format that is based on well articulated proportional and rhythmic systems. The architectural language underlying the Palladian villa plan designs was the focus of some pioneering studies of the shape grammars (e.g., [Stiny and Mitchell 1978]). Although the languages of all Palladian designs in two- and three-dimensional representations are significant for the visual studies and architectural research, the focus of this paper will be on a new representation of the morphology of Palladian palazzo façade designs.

The palazzos in this study sample consist of nine buildings. These include the houses of:

1. Signor Floriano Antonini in Udine (P1)
2. Count Valerio Chiericati in Vicenza (P2)
3. Count Iseppo De Porti in Vicenza (P3)
4. Count Giovanni Battista Della Torre in Verona (P4)
5. Count Ottavio De Thiene in Vicenza (P5)
6. Counts Valmarana in Vicenza (P6)
7. Monsignor Paolo Almerico in Vicenza (P7)
8. Signor Giulio Capra in Vicenza (P8)
9. Count Montano Barbarano in Vicenza (P9)

In the following sections, the string development and recognition systems introduced in the previous sections will be applied to a selected set of the façades of Palladio's palazzos to highlight their formal language. As such, façades of the above listed nine houses will be analyzed according to their lexical and syntactic structures, and then encoded in the form of FSA representations.

The Palladian Façade Language

The architectural elements underlying the Palladian palazzo designs reveal clear ingredients of a formal language that can be assembled into a finite alphabet and a finite set of compositional rules (e.g., rules for symmetry and rhythm). The major concern of this section will be to highlight the morphology of this language as enunciated in Palladio's palazzo designs, as well as the process of composing the planar visible configurations of these designs as manifested in a selected set of their various façades.

The main vocabulary elements shared by the façades of Palladian palazzos consist of the wall units (solid, with door, with window, with two windows, with door and window, or negative which represents a space between walls or columns), columns (Doric, Ionic, Corinthian, and Composite), pediments, entablatures (that vary according to their cornice, frieze, and architrave details), roofs (pitched or domed), and sculptural units. The hierarchy of the vocabulary elements of the Palladian façade language is illustrated in fig. 2. The sculptural units will not be discussed in the morphology analysis as they are out of the scope of this article. Among the remaining vocabulary elements, the emphasis will be on wall, opening and column units. Examples of the architectural vocabulary components of the Palladian façades are shown in fig. 3. These components are the ones referred to in the FSA encoding of the front façade of Palazzo Antonini, which is detailed in the following section. The Palladian vocabulary set consists of:

W: wall unit:

- W0: space between walls or columns.
- Ws: solid wall without openings.
- W1: wall with window. It has the variations W11 (example in P1), W12 (example in P3) and W13 (example in P3).
- W2: wall with two windows. It has the variations W21 (example in P1) and W22 (example in P6).
- W3: wall with door. It has the variations W31 (example in P2) and W32 (example in P4).
- W4: wall with door and window. It has the variations W41 (example in P1) and W42 (example in P9).

C: column unit:

- C1: Ionic (example in P1).
- C2: Composite (example in P1).
- C3: Doric (example in P2).
- C4: Corinthian (example in P8).

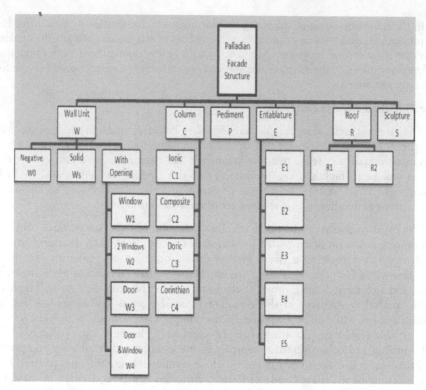

Fig. 2. A hierarchical chart of some vocabulary elements of the Palladian palazzo façades

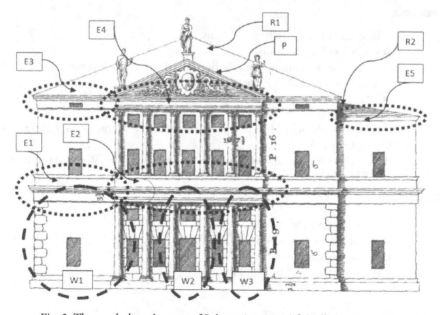

Fig. 3. The vocabulary elements of Palazzo Antonini (cf. [Palladio 1997: 80])

Encoding the Palladian Façades by Finite State Automata

In this section, selected front and courtyard façades of the aforementioned nine houses will be represented by strings of symbols. For each palazzo, a finite state automaton diagram will be illustrated. Both forms of representation, the symbolic and the graphical, will describe the Palladian language of façade compositions. The first palazzo will be described in more details than the others. All representations introduced for the first palazzo may be applied to all of the others. However, for space management the representations of the remaining eight palazzos will be kept to minimum.

1. The Palazzo of Signor Floriano Antonini in Udine (P1):

In terms of its compositional language, the façade of Palazzo Antonini consists of five major horizontal levels (see fig. 4). These are: L1) the first floor elevation, L2) the decorative level that separates the walls and columns of the first floor from the second floor level, L3) the second floor level, L4) the decorative level that crowns the second floor walls and separates them from the roof level, and finally L5) the roof level in which pitched masses close the spaces underneath.

The composition of each façade level can be thought of as an input tape that is read from left to right, in order to encode the level as a string of symbols. The process of composing each string is similar to the string processing sequence for regular languages using FSAs, discussed earlier in this article. The initial state corresponds to the leftmost edge of the building, and transitions between states occur as the FSA reads the symbols that correspond to the architectural elements. The process of composing the strings of each level in Palladian palazzo façades moves through a maximum of seven states, Q0–Q6, shown in fig. 4. The FSA for Palazzo Antonini's façade is illustrated in fig. 5.

For example, for the first level, L1, the procedure starts with the initial state Q0, adds a wall unit with window (W11) and moves to state Q1. Upon the addition of a column unit (C1), the automaton remains in state Q1, but upon the addition of a wall unit with two windows (W21), it transitions to Q2. Similarly, a column unit keeps the state at Q2 whereas a wall unit with two windows moves it back to Q1. The FSA moves between Q1 and Q2, and vice versa, adding a column and a wall with two windows each time, until half of the columns are processed. The Q1–Q2 sequence thus adds three columns and two wall units. A wall with door and window (W41) is then added through the movement from state Q1 to state Q3. Cycles between Q3 to Q4 are directly analogous to cycles between Q1 and Q2, since they are mirror images around a vertical axis that passes through the center of the door. Thus the Q3–Q4 sequence also adds three columns and two wall units. Upon the addition of a wall with one window, the automaton moves from state Q3 to state Q5 to conclude the symmetric part of the façade. Finally, the rightmost asymmetric wall (W11) is added. This last step transforms the FSA from state Q5 to the final state Q6.

With some change in the input symbols, this FSA can be applied identically in level L3. The major difference in the input symbols lies in the column order, which is Ionic (C1) in L1 and Composite (C2) in L3. In the ornamental levels L2 and L4, states Q2 and Q4 can be dropped. Similarly, in the roofing level L5, the states of Q1 through Q4 are not needed.

The strings for each level of the Palazzo Antonini façade are:

L1 = W11 C1 W21 C1 W21 C1 W41 C1 W21 C1 W21 C1 W11 W11

L2 = E1 E2 E1 E1 (these elements vary according to their level of detail in decoration)

L3 = W11 C2 W21 C2 W21 C2 W41 C2 W21 C2 W21 C2 W11 W11

L4 = E3 E4 E3 E5

L5= R1 R2 (these elements vary according to their shapes)

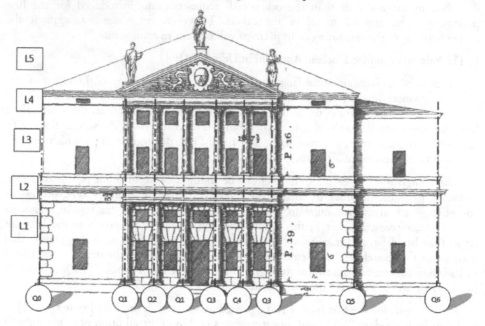

Fig. 4. The main levels (L1–L5) and states (Q0–Q6) of the front façade of Palazzo Antonini (cf. [Palladio 1997: 80]). States are indicated on the façade just before they transition to the next state, which results in Q1–Q4 being positioned at the columns

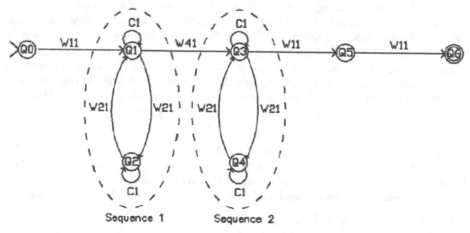

Fig. 5. The finite state automaton for the front façade of Palazzo Antonini

---

The graphic interpretations of the symbols in the previous strings are illustrated in fig. 3. For the first level L1, the transitions are:

Q0→Q1→Q1→Q2→Q2→Q1→Q1→Q3→Q3→Q4→ Q4 →Q3→ Q3→Q5 →Q6.

The input/output values for the transition function are shown in Table 2. As in Table 1, each row corresponds to an input/output value of the transition function, where the first two columns are the input and the last column is the output.

The FSA of the front façade of Palazzo Antonini is formally represented by M = (K, Σ, δ, S, F), where:

K is the finite set of states = {Q0, Q1, Q2, Q3, Q4, Q5, Q6}
Σ is the alphabet = {Ø, W11, W21, W41, C1, C2, E1, E2, E3, E4, E5, P, R1, R2}
S ∈ K is the initial state = Q0
F is the set of final states (subset of K) = {Q6}
δ is the transition function (from K x Σ to K).

For example, the transition functions of level L1 are (see Table 2):

δ(Q0,W11) = Q1,
δ(Q1, C1) = Q1,
δ(Q1,W21) = Q2,
δ(Q1,W41) = Q3,
δ(Q2,C1) = Q2,
δ(Q2,W21) = Q1,
δ(Q3,C1) = Q3,
δ(Q3,W21) = Q4,
δ(Q3,W11) = Q5,
δ(Q4,C1) = Q4,
δ(Q4,W21) = Q3,
δ(Q5,W11) = Q6.

q	σ	δ(q, σ)
Q0	W11	Q1
Q1	C1	Q1
Q1	W21	Q2
Q1	W41	Q3
Q2	C1	Q2
Q2	W21	Q1
Q3	C1	Q3
Q3	W21	Q4
Q3	W11	Q5
Q4	C1	Q4
Q4	W21	Q3
Q5	W11	Q6

Table 2 (right). The table of input/output values for the transition function of level L1 of Palazzo Antonini's front façade

The same transition function works for level L3, with C1 replaced by C2. Of course, the transition functions for levels L2, L4, and L5 would be considerably simpler.

All the representations produced for the Palazzo Antonini façade can be applied to the remaining eight palazzos. However, formulas and graphs for only one level of each of these palazzos will be demonstrated in the following sections. The number and locations of states (Q0-Q6) may vary for the following houses, as can the assignment of initial and final states. However, commonalities among these will be discussed later in this article.

2. The Palazzo of Count Valerio Chiericati in Vicenza (P2):

This house is built upon a piazza in Vicenza and raised five feet above the ground level. The order on the first level is Doric and on the second is Ionic. The state locations and additional vocabulary elements are illustrated in fig. 6. The level analyzed for this façade is L1. Note that the fourth column from the left (only half of which is visible) is located in a different plane (its axis is shifted behind the main plane of all others), and so it is not included in the analysis (the same is true for the fourth column from the right).

The FSA in fig. 7 represents L1 of this palazzo. Columns of this level are of the Doric (C3) order. The same FSA applies to L3 with a change of column style into Ionic (C1).

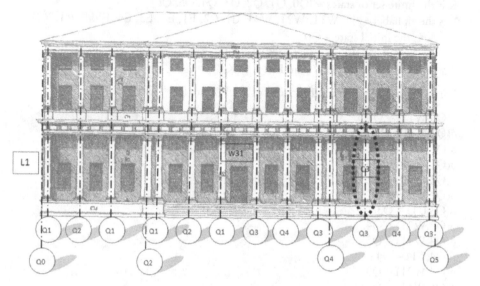

Fig. 6. The state assignment of the front façade of Palazzo Chiericati (cf. [Palladio 1997: 83])

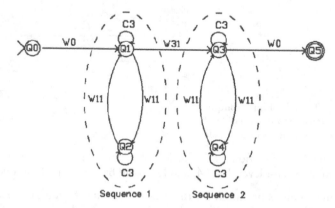

Fig. 7. The FSA of the front façade of Palazzo Chiericati

3. The Palazzo of Count Iseppo De Porti in Vicenza (P3):

This house has two main façades, each of which includes an entrance from a public street. The façade analyzed here is the front one, half of which is shown in fig. 8. The following automaton (fig. 9) represents L3 of this palazzo. Columns of this level are of the Ionic order. Ws stands for a solid wall. W12 and W13 symbolize wall with window units, the former being crowned with an ornamented triangle, and with an ornamented arch in the latter. Note that the FSA is nondeterministic: state Q2 together with symbol W13 can lead to either state Q1 or state Q4.

Fig. 8. The state and vocabulary assignment of one half of the façade of Palazzo De Porti (cf. [Palladio 1997: 85])

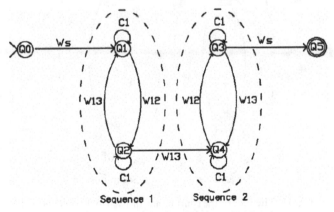

Fig. 9. The FSA of the façade Palazzo De Porti

4. The Palazzo of Count Giovanni Battista Della Torre in Verona (P4):

This house in Verona has a long and narrow rectangular plan, for which Palladio provides two interior courtyards. The house faces a principal street with one of its narrow sides. The façade Palladio shows is the one resulting from a section through the inner courtyards, illustrated in fig. 10.

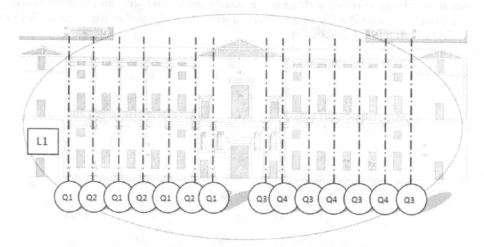

Fig. 10. The state assignment of the court façade if Palazzo Torre (cf. [Palladio 1997: 87])

The following FSA (fig. 11) represents L1 of this palazzo. Columns of this level are of the Ionic order (their order is Composite in L3). Since transition from state Q1 with the symbol W0 could be to either Q2 or Q3, this FSA is nondeterministic (as are those for P5, P7, and P8 below).

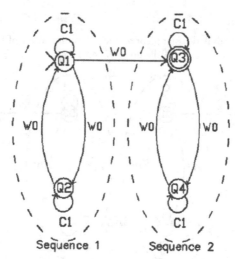

Fig. 11. The FSA of the court façade of Palazzo Torre (Verona)

---

5. The Palazzo of Count Ottavio De Thiene in Vicenza (P5):

This house is located in the middle of Vicenza city near the Piazza. As in the previous palazzo, the façade of interest here is the one facing the inner courtyard that is illustrated in fig. 12.

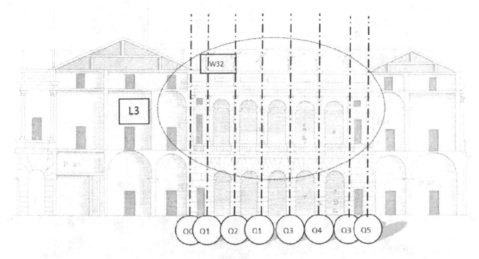

Fig. 12. The state and level assignment of the court façade of Palazzo Thiene (cf. [Palladio 1997: 89])

The following FSA (fig. 13) represents L3 of the façade facing the courtyard. The courtyard is encompassed from all sides with loggias of pilasters. The order of court's L3 columns is Composite. W41 symbolizes a wall with door and window unit, and W32 symbolizes a wall with arched opening in it.

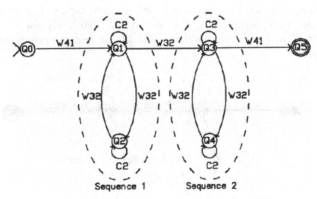

Fig. 13. The FSA of the court façade of Palazzo Thiene

6. The Palazzo of Counts Valmarana in Vicenza (P6):

This house consists of two parts divided by a middle court. The façade studied here is the front one (fig. 14).

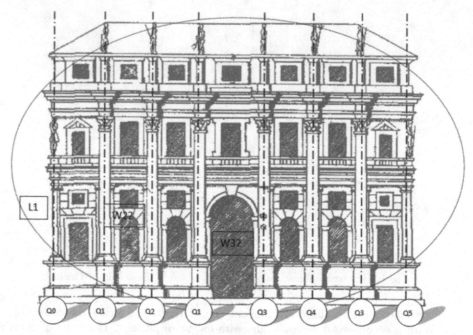

Fig. 14. The state and level assignment of the court façade of Palazzo Valmarana (cf. [Palladio 1997: 92])

The following FSA (fig. 15) represents L1 of the front façade. The scale of C2 (column of Composite order) is multiplied by 2 to extend over two levels (for six major columns). W22 symbolizes a wall with two windows, one square and the other arched. W32 is an arched gateway.

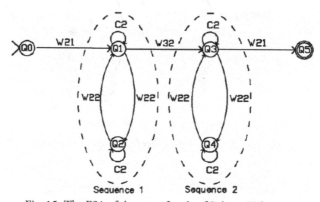

Fig. 15. The FSA of the court façade of Palazzo Valmarana

---

7. Monsignor Paolo Almerico City Villa in Vicenza (Villa Rotonda) (P7):

Although this famous building is a country villa upon a hill close to Vicenza, Palladio classified it with city houses because of its site's proximity to the city [Palladio 1997: 94]. The FSA of this example is developed for its typical front portico illustrated in fig. 16.

The following FSA (fig. 17) represents the loggia at the front of each side of the symmetric four façades.

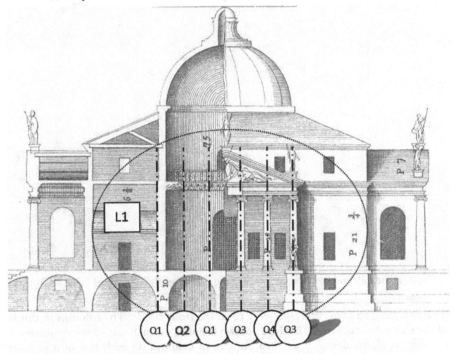

Fig. 16. The state assignment of the Façade of Villa Rotonda (cf. [Palladio 1997: 95])

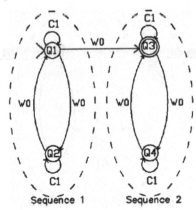

Fig. 17. The FSA of the Façade of Villa Rotonda

8. The Palazzo of Signor Giulio Capra in Vicenza (P8):

This house is located on the principal street of Vicenza. The shape of its site is irregular in plan. However, its façades exhibit a language that is consistent with the others analyzed so far (fig. 18).

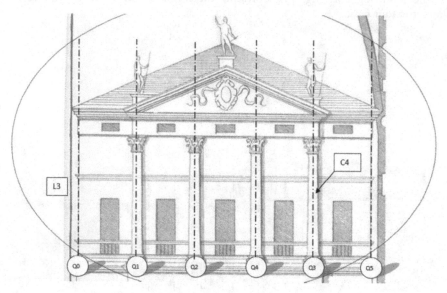

Fig. 18. The state and level assignment of the upper floors of the façade of Palazzo Capra (cf. [Palladio 1997: 97])

The following FSA (fig. 19) represents L3 of the front façade. The columns in this level are of the Corinthian order (C4). There are two columns on each side of the symmetry axis. As a result (in all façades with an even number of columns on each half of a symmetric façade), the FSA moves from Q2 to Q4 instead of Q1 to Q3 (with an odd number of columns on each half).

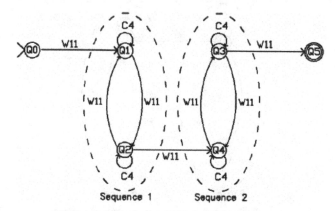

Fig. 19. The FSA of the façade of Palazzo Capra

9. The Palazzo of Count Montano Barbarano in Vicenza (P9):

In this façade, columns of the Composite order (C2) are scaled to extend over two stories (fig. 20). In this case L1 and L3 can be combined in one FSA representation (fig. 21). W42 symbolizes a wall unit with rectangular window and arched gateway.

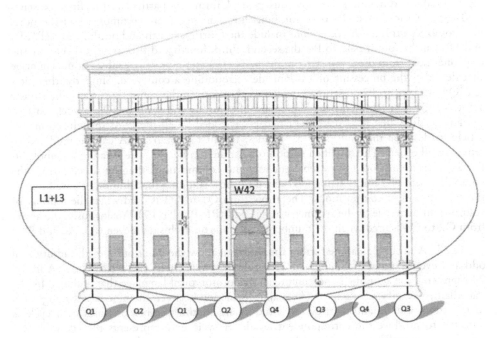

Fig. 20. The state assignment of Palazzo Barbarano (cf. [Palladio 1997: 98])

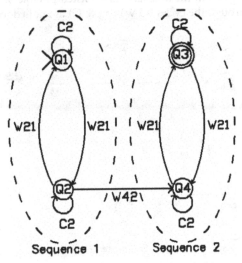

Fig. 21. The FSA of the façade of Palazzo Barbarano

The Palladian Palazzo Façade Language

By comparing the nine façades via their graphic and string representations, the main structure of the FSA of Palazzo Antonini (P1) as illustrated in figs. 4 and 5 seems to represent the most comprehensive form among all others. All other façades can be represented by different subsets of this general form. In particular, the first six states (Q0–Q5) of the FSA of Palazzo Antonini shown in fig. 5 are common to all the other palazzos except three. The exceptions include the fourth, seventh and ninth houses (P4, P7 and P9). In the fourth palazzo P4, the second, third, fourth and fifth states Q1–Q4 are the only ones kept from the seven states of fig. 5. These states represent a subset of the language that describes the processing of a colonnade surrounding a courtyard. Similarly, the states Q1–Q4 are kept in P7 to represent columns that surround a central gateway. The similar subset of Q1–Q4 represents the alternation of columns and walls in P9. As such, to find the general FSA that recognizes the languages encoded by all the FSAs introduced in this article to describe the nine palazzos' façades, the first palazzo's FSA can be modified to represent all others. The modified version is illustrated in fig. 22. The W and C symbols (in fig. 22) that appear on arrows represent wall and column units that are selected from the vocabulary set. The empty string Ø is added to allow skipping the addition of a wall unit at the start or end of compositions. The arrow from Q1 to Q3 is used when the number of columns on each side of the symmetry axis is odd (P1–P2, P4–P7). Analogously, the arrow from Q2 to Q4 is used when the number of columns on each side is even (P3, P8, and P9).

The FSA in fig. 22 can be further simplified to allow the description of the addition of odd and even number of columns on each side of symmetry in one cycle. The FSA in fig. 23 represents the simplified and comprehensive prototypical language that is shared by the Palladian palazzo façade designs. The major distinction between the nine façade designs lies in the transition functions in cycles 1 and 2 and in the number of times each cycle is repeated to produce the correspondent façade. A cycle here represents the addition of a column and a wall (or space), and forms a part of the sequence described in the previous FSAs. The numbers of cycles in the nine façades are listed in table 3. These values represent the number of loops through the cycles (C W) or (W C) as applied for each house.

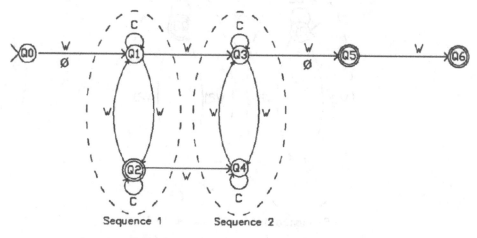

Fig. 22 (above). The generic FSA of Palladian Palazzo façades

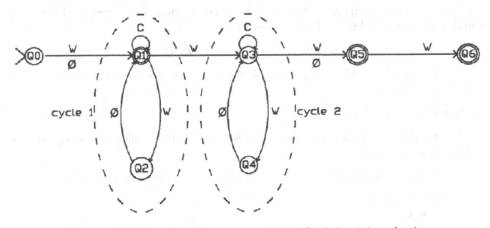

Fig. 23 (below). The simplified prototypical FSA of Palladian Palazzo façades

House	CYCLE 1 (CW) $\delta((Q1, C) \to Q1 \to (Q1, W) \to Q2))$ m	CYCLE 2 (WC) $\delta((Q3, W) \to Q4 \to (Q3, C) \to Q3))$ n
P1	2	2
P2	6	6
P3	3	3
P4	6	6
P5	2	2
P6	2	2
P7	2	2
P8	1	1
P9	3	3

Table 3. The transition functions of the major cycles of the prototypical Palladian palazzo façades FSA of fig. 23. In the transition function for CYCLE 1, Q1 and Q2 can be interchanged; for CYCLE 2, Q3 and Q4 can be interchanged

The common language of Palladian palazzo façades is generated by the nondeterministic FSA M = (K, Σ, δ, S, F), where

K is the finite set of states = {Q0, Q1, Q2, Q3, Q4, Q5, Q6}

Σ is the alphabet = {Ø, W, C}, where W stands for a wall unit, Ø for an empty shape and C for a column

S ∈ K is the initial state = Q0

F is the set of final states (subset of K) = {Q1, Q5, Q6}; Q1 allows the recognition of colonnade compositions

δ is the transition function (from K x Σ to K), comprised of the transitions:

δ(Q0,W) = Q1,
δ(Q0,Ø) = Q1,
δ(Q1,C) = Q1,
δ(Q1,W) = Q2,
δ(Q1,W) = Q3,
δ(Q2,Ø) = Q1,
δ(Q3,C) = Q3,
δ(Q3,W) = Q4,
δ(Q3,W) = Q5,
δ(Q4,Ø) = Q3,
δ(Q5,W) = Q6,

Given that:

W: a wall (solid, with openings, or negative) unit

C: column of a given order

Ø: empty string (allows change of state without processing a façade element).

All the words that can be composed for Palladian palazzo façades are variations of the general form:

$$W^a \, (C \, W)^m \, C \, W \, C \, (W \, C)^n \, W^b$$

for some integers $a, b, m, n \geq 0$.

The number of cycles, m and n, for each of the nine façades analyzed, are listed in Table 3. Note that $m=n$ in all façades. As such, the language becomes:

$$W^a \, (C \, W)^n \, C \, W \, C \, (W \, C)^n \, W^b$$

(for the nine façades the integers satisfy $0 \leq a \leq 1$, $1 \leq n \leq 6$, $0 \leq b \leq 2$).

The middle part of the string, C W C, represents the central part of each façade through which the axis of symmetry passes. This part is always a wall (or space) that is surrounded by two columns. It is, in turn, surrounded by an alternation of walls and columns from both sides. For a colonnade, W is replaced with W0 (a space).

As the prototypical automaton illustrated in fig. 23 suggests, the first symbol of a Palladian façade string consists of a wall unit that is applied zero or more times (zero meaning that this unit can be skipped in some façades to start the façade strip with a column), followed by a loop that consists of an alternation of a column and a wall unit. The string has always a central word CWC that represents the main gateway unit. Then the first-half sub-string is mirrored around its central word to produce a symmetric sub-string. In P1, an additional unit is added to make the string asymmetric.

The FSA-based representations of the nine Palladian façades as introduced in this article help emphasize the common language that unifies all of them and may help produce new Palladian façades that conform to the same vocabulary elements and process sequence. The illustrated representations demonstrate also that a simple regular language recognition device can be applied to explain the processing of a physical object's components such as those constituting architectural façades. Although such an application is not suggested to be intuitive, it proposes more than an intellectual exercise. Such an added layer of interpretation can highlight the commonalities of different structures. It can classify collections of art products and encode them in short strings that facilitate their comparisons and management.

The pictorial prototypical representation of the FSA (fig. 23) can be further developed to automatically model elevations by feeding series of different symbols iterated as many times as needed. These symbols can then be replaced by graphical units to draw their correspondent building elevations by computer. For instance, such graphical units may be thought of as pre-stored blocks in AutoCAD or library cells in 3DMax. Each block or cell may replace a designated symbol as requested by the user. Furthermore, each symbol may tag a command that linearly adds a graphic unit that may be scaled and assigned to a style as needed.

Conclusion

A selected set of the façades of nine city houses that are designed by Palladio has, in this article, been analyzed to externalize their underlying common formal language. By the isolation of each level of the façade at hand, a word in the architectural language can be conceived as a linear string that can be sequentially processed one symbol at a time. As such, the selected set has been represented by a string recognition device that is known as a finite state automaton. The concepts underlying this automaton are borrowed from formal language studies and their computational applications. The proposed string-based representations are formulated in terms of symbolic formulas and graphical directed diagrams. A prototypical automaton is concluded for the Palladian façades of the study sample instances. It represents the formal language common to Palladian palazzo façades. Upon feeding the right symbols and cycling as many times as ordered by the user, the automaton can regenerate the existing façades and can propose new emergent façades that belong to the same style language. The finite state automata representation is demonstrated to help add a different layer of interpretation to the commonalities and differences of the façade structures and their underlying language. Similarly developed prototypes can be derived to represent other Palladian building types which can be compared to the ones analyzed in this article. These prototypes can be further developed to compare different façade languages and to automatically generate existing and emergent façades.

References

AHO, A.V., R. SETHI, and J.D. ULLMAN. 1986. *Compilers, Principles, Techniques, and Tools*. Reading: Addison-Wesley Publishing Company.

AHO, A.V. and J.D. ULLMAN. 1972. *The Theory of Parsing, Translation, and Compiling*. Englewood Cliffs: Prentice-Hall, Inc.

ARNHEIM, R. 1960. *Visual Thinking*. Berkeley: The University of California Press.

———. 1977. *The Dynamics of Architectural Form*. Berkeley: The University of California Press.

BARWISE, J. 1996. *Logical Reasoning with Diagrams*. New York: Oxford University Press.

BARWISE, J. and J. ETCHEMENDY. 2001. *Turing's World 3.0: An Introduction to computability theory*. Stanford: CSLI Publications.

CAWS, P. 1988. *Structuralism: The Art of the Intelligible*. Atlantic Highlands: Humanities Press International, Inc.

CHOMSKY, N. 1964. *Current Issues in Linguistic Theories*. Netherlands: Mounton and Co. Printers.

DING, L. and J.S. GERO. 2001. The Emergence of the representation of Style in Design. *Environment and Planning 'B': Planning and Design* 12: 59-73.

EILOUTI, B. H. 2005. The Representation of Design Sequence by Three-Dimensional Finite State Automata. Pp. 707-731 in *The International Institute of Informatics and Systemics*, D. Zinn (ed.). Orlando, FL.

———. 2007. A Spatial Development of a String Processing Tool for Encoding Architectural Design Processing. *Art, Design and Communication in Higher Education* 6, 1: 57-71.

EILOUTI, B. H. and E.G. VAKALO. 1999. Finite State Automata as Form-Generation and Visualization Tools. Pp. 220-224 in *Information Visualization IV 99*, E. Banissi and F. Khosrowshahi, eds. Los Alamitos: IEEE Computer Society.

FRANKEL, F. 2005. Translating Science into Pictures: A Powerful Learning Tool. Pp. 155-158 in *Invention and Impact: Building Excellence in Undergraduate Science, Technology, Engineering, and Mathematics (STEM) Education*. AAAS Press.

HARRISON, M. A. 1978. *Introduction to Formal Language Theory*. Massachusetts: Addison-Wesley Publishing Company.

HERSEY, G. and R. FREEMAN. 1992. *Possible Palladian Villas (Plus a Few Instructively Impossible Ones)*. Cambridge, MA: MIT Press.

HOPCROFT, J. E. and J.D. ULLMAN. 1979. *Introduction to Automata Theory, Languages, and Computation*. Reading: Addison-Wesley Publishing Company.

KOUTAMANIS, A. 1994. The Future of Visual Design Representations in Architecture. Pp. 76–98 in *Automation Based Creative Design: Research and Perspectives*. Tzonis and White, eds. New York: Elsevier Science BV.

LEWIS, H. R. and C.H. PAPADIMITRIOU. 1981. *Elements of the Theory of Computation*. Englewood Cliffs: Prentice Hall.

LINZ, P. 1997. *An Introduction to Formal Languages and Automata*. Massachusetts: Jones and Bartlett Publishers.

MARR, D. 1982. *Vision: A Computational Investigation Into the Human Representation and Processing of Visual Information*. San Francisco: W. H. Freeman.

MCGRATH, M. B. and J.R. BROWN. 2005. Visual Learning for Science and Engineering. *IEEE Computer Graphics and Applications* 25, 5: 56-63.

MITCHELL, W. J. 1990. *The Logic of Architecture: Design, Computation and Cognition*. Cambridge, MA: MIT Press.

NAVINCHANDRA, D. 1990. *Exploration and Innovation in Design*. New York: Springer-Verlag.

PALLADIO, A. 1997. *The Four Books of Architecture*. Robert Tavernor and Richard Schofield, trans. Cambridge MA: MIT Press.

REVESZ, G. E. 1983. *Introduction to Formal Languages*. New York: McGraw-Hill.

SALOMAA, A. 1973. *Formal Languages*. New York: Academic Press.

SHEPPARD, S. 1989. *Visual Simulation: A User's Guide for Architects, Engineers, and Planners*. New York: Van Nostrand Reinhold.

STINY, G. 1975. *Pictorial and Formal Aspects of Shape and Shape Grammars*. Basel: Birkhäuser.
———. 1994. The Pedagogical Grammar. Pp. 129-139 in *Automation Based Creative Design: Research and Perspectives*. Tzonis and White, eds. New York: Elsevier Science BV.
STINY, G. and W.J. MITCHELL. 1978. The Palladian Grammar. *Environment and Planning 'B': Planning and Design* 5: 5-18.
TAPIA, M. 1999. A visual implementation of a shape grammar system. *Environment and Planning 'B': Planning and Design* 26: 59-73.
WOJTOWICZ, J. and W. FAWCETT. 1986. *Architecture: Formal Approach*. New York: St. Martin's Press.

About the author

Buthayna H. Eilouti is Associate Professor in the department of Architectural Engineering in Jordan University of Science and Technology. She earned a Ph.D., M.Sc. and M.Arch. degrees in Architecture from the University of Michigan, Ann Arbor, USA. Her research interests include Design Studies, Design Mathematics and Computing, Design Pedagogy, Visual Studies, Shape Grammar, Information Visualization, and Islamic Architecture.

Tomás García-Salgado

Facultad de Arquitectura, UNAM
Ciudad Universitaria
Coyocán, México
tsalgado@perspectivegeometry.com

Keywords: Andrea Palladio, Villa
Rotonda, Design analysis,
geometric analysis, perspective

Research

A Perspective Analysis of the Proportions of Palladio's Villa Rotonda: Making the Invisible Visible

Abstract. Because of differences between Palladio's architecture as built and the ideal architecture represented in the *Quattro libri*, many analyses have been performed in order to bring to light the proportions that underlie the beauty of the architecture. This present paper proposes a method of analysis based on perspective grids laid out on photographs to reveal how perspective is used to heighten the spectator's perception of the forms.

Introduction

Many architects and scholars agree that the Villa Rotonda epitomizes the architectural ideals that Palladio set forth in his *Quattro libri* [1570]. Palladio's villas have been profusely copied, but often cheerlessly or with no understanding of their cultural and temporal context. In England, Palladio's architectural style was adopted after the English translation of his book and soon after the same happened in America. The fact that the drawing of the Rotonda published in the *Quattro Libri* does not correspond exactly with its actual built conditions has motivated the elaboration of detailed drawings for this building, such as those by Bertotti-Scamozzi [1776-83]. More recently, Semenzato has obtained more accurate plans by using internal polygonals [1990]. Artistic drawings of all kinds have also been made, especially in watercolors, such as those of Giovanni Giaconi [2003]. Models assisted by computer technology have also been produced, as those developed for all of Palladio's villas by Lawrence Sass at MIT [2001]. However, I am convinced that superimposing grids on photographs of the villa could lead to a new approach towards finding out its proportions as they are perceived in perspective, and thus tell us something more about its singular composition.

Palladio's drawings versus the actual building

Palladio's woodcut engravings for the *Quattro libri* [1570] were probably intended to express his architectural ideals, not the buildings as they were actually built. As Sass pointed out, "Palladio created the drawings and text in the *Four Books of Architecture* after the buildings were built" [2001: 6]. The Rotonda was begun in 1550, whereas the first edition of the *Quattro libri* was published in 1570, meaning one of two things: either he chose to publish an idealized drawing instead of a real one —if such a "real" drawing ever existed—, or he deliberately prepared an ideal drawing aimed at presenting his canons. Curiously, the history of the ownership of the Rotonda's estate is known down to the slightest detail, but little or nothing is known regarding the villa's proportions. Whatever the answer may be, we still have to deal with some dissimilar proportions between the ideal drawing and the actual building. For instance, in the *Quattro libri*, the northeast and southwest vestibules are shown with a width of $6f$ (f being one Vicentine foot), and those of the northwest and southeast with $7\frac{1}{2}f$ (as deduced in fig. 1), whereas the latest surveys indicate $10\frac{1}{2}$-$11f$ in reality. This dissimilarity changes the proportions of the main rooms from $15 \times 26f$, to 15×24-$24\frac{1}{4}f$. How do we explain that?

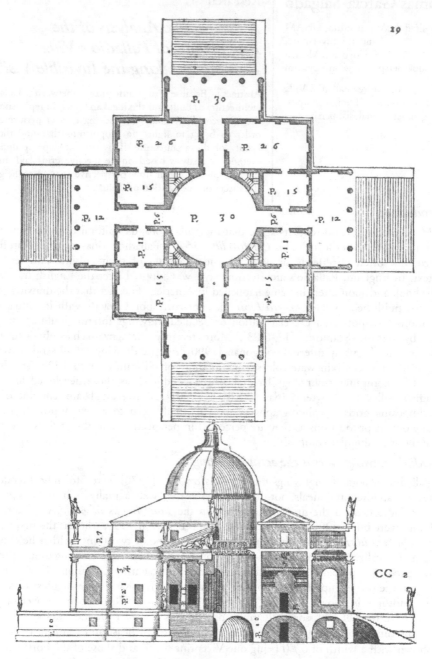

Fig. 1. Palladio's drawing of the Rotonda in the *Quattro Libri Dell'Architettura* (Plate of page 19, *Libro Secondo*). In the elevation, the two windows of the loggia at the *piano nobile* indicate this is a lateral view of the plan, not the frontal one as the drawing arrangement suggests

In plate number 19 in the *Quattro libri* (fig. 1), the thickness of the walls and the height of the central vault are not indicated; only the proportions of the rooms are given. The height of the columns of the loggias is indicated, but their diameters, which is important for verifying the 12 x 30 *f* proportions of the loggias, as we will see shortly. In the same plate, we see how the central intercolumniation appears slightly larger than the others, suggesting an arrangement in accordance with the precepts of eustyle and diastyle. Obviously, it would take much more than a single plate to accurately depict the architectural elements of the loggias alone, not to mention the many more needed to render the whole building. It seems that Palladio was more interested in justifying the motivation for his design for the loggias[1] than he was in explaining the geometrical layout in detail.

Palladio devoted his first book to a detailed description of the classical styles (*toscano, dorico, ionico, corinthio, composito*), so perhaps for this reason he did not feel compelled to do the same for his own works; after all they can be inferred from the orders. There are no descriptions at all of how his buildings were built, nor are working drawing for this purpose included, if any were indeed ever elaborated. It seems obvious that the stonemasons required some sort of drawings to carve the capitals and the stone frames for the doors and windows, to shape the entasis of the columns, and the moldings of the pediments. In other words, the challenging question here is how Palladio managed the geometry of his buildings during construction, to which we have no answer. It seems that Palladio wanted us to learn the ideals from his treatise and the reality from his works, leaving our imagination to bridge the gap between his theory and practice. The perspective analysis I am proposing here intends to bridge, in part, that gap.

Squares within squares

The geometry of the *square* seems to rule the configuration of the Rotonda's plan, since several imaginary squares can be identified in a concentric position alternating with other squares, which are rotated 45° from the main plan. I discussed this particular feature in a previous article, showing how these squares relate to one another, from the stairways to the nucleus of the building [2004]. Based on this approach, here I will describe its spatial meaning to link it with the discussion in the next section. Thus, to proceed in an orderly fashion, I will refer first to the analysis of the plan, and later to its volumetric interpretation in perspective.

As can be noticed in fig. 2 (*A* through *H*), the arrangement of the plan has mirror symmetry along both the longitudinal and the transversal axes. We had to assume a hypothetical value for the thickness of the walls to redraw the plan, since Palladio's drawing lacks this data. Lawrence Sass has estimated the wall thicknesses to be 18 inches at the *piano nobile* [Sass n.d.: 8]; according to my calculations equals approximately 1-1/4 *f*, almost the same as the dimension estimated by Sass, although the surrounding walls could be thicker, as much as 1½ *f.*

The sequence of the squares *A-B-C* can help us understand the spatial integration between the villa and its environment (see fig. 2, *A*, *B*, *C*). We will begin by demarcating the ground plan at the four stairways as square *A*. This square has a spatial meaning that is not easy to grasp until one surveys the villa from different vantage points, wondering why it looks bigger than it actually is. This is followed by Square B surrounding the projecting loggias of the building that characterize its unique formal expression. Square *C* integrates the composition of the whole toward which the invisible squares *A* and *B* converges.

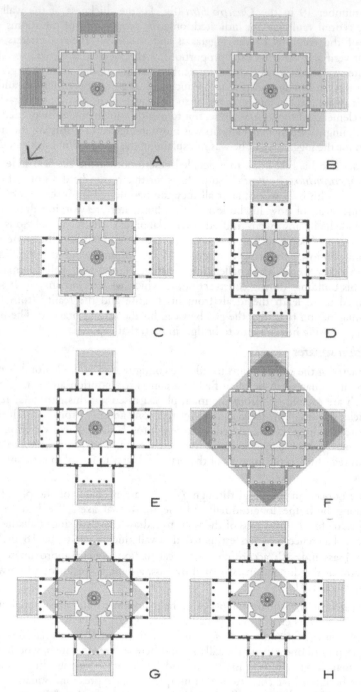

Fig. 2. (*A* through *H*). Squares within squares. Drawings by Ambar Hernández and author

On the interior, square D is formed by a continuous perimeter passing throughout all the doors of the rooms that surround the central hall. This feature of spatial continuity could have been helpful during construction for checking the position of the inner walls by simply stretching a cord along the perimeter of square D. Square D crosses all the chambers either at the center of their doors, or at one side of them; this seems to have been planned from the beginning for this purpose at the *piano terra*. In its turn, square E holds the transition between the twelve rooms (including the halls) and the circular hall, besides the stair system connecting all the floors vertically.

Now, the first square rotated by 45°, F, is obtained by inscribing the four corners of the building, passing from the plinth of the extreme columns of the loggias until the center of the stairways is reached. Standing at the center of the first step of a stairway, the relationship among squares A, B, and C makes sense, since no other loggia can be seen in it but the one in front of us. Another rotated square G can be traced from the first step of the loggias running across the main chambers, inscribing at the same time square D. Finally, square H joins at the thresholds of the four main entrances, inscribing the nucleus of the house while crossing at one side of the halls' doors. Some other interesting relationships can also be found by extending both diagonals of any of the main chambers, the same being true for the diagonals of the small chambers.

The significance of "the whole" and "the parts" in Palladio's architecture becomes clearer and more comprehensible when it is perceived in perspective, as in the hyperrealistic fig. 3. Here, an imaginary geometry of volumes A, B, and C, on a photograph taken from the north, allow us to visualize why the villa looks bigger than it actually is, as we said before. Certainly, after walking for a while in the gardens, one can perceive the building within a virtual volume, extending its dimensions beyond its walls, and vice versa, and how the landscape embraces it gradually.

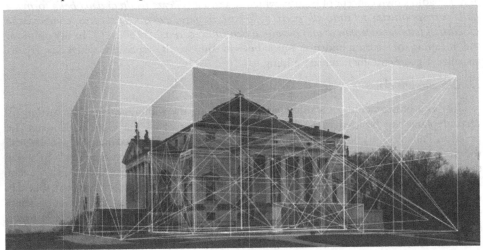

Fig. 3. A hyperrealistic representation of the virtual squares A and C. The reader will be able to imagine where should square B lay —between A and C. Photograph and superimposed drawing by the author

Laying out grids in perspective ...

Now, to analyze the volumetric composition of the building in perspective, we will lay out some virtual grids on its façades, combining squares and rectangles, and see that several of their diagonals make explicit the proportions of the windows, doors, and the loggias. Of course, I am not suggesting here that Palladio literally drew grids on the façades in order to place the elements. Those who have been involved in architectural practice know how geometry is applied while building; how dimensions and proportions are as easily adjusted by moving cord-lines in the field, as lines are moved with an straightedge on a drawing table.

According to the *Quattro libri*, the original proportions of the Rotonda are as follows: in plan, 12 x 30 f for the loggias, 15 x 26 f for the main rooms, 11 x 15 f for the small rooms, 6 f for the access halls, and a $r = 30 f$ for the circular room. In elevation: the height of the columns is 18 f, that of the main rooms is $21^1/_2$ f, the attic is 7 f, and there are 10 f from the *piano terra* to the *piano nobile*. The question now is how to visualize these proportions in three dimensions.

We have already established that Palladio's drawing does not provide enough information about the elements of the façades. This notwithstanding, he must have decided how to manage its proportions *in situ* during construction, somehow applying a geometrical procedure (at least, this is my hypothesis). I cannot otherwise explain the many coincidentally related elements that become evident when a virtual grid is superimposed on the façades – in this case by means of photographs. In a way, these grids can reconstruct the design process of placing the elements of the façades in reverse, revealing their geometrical proportions. Thus, windows, doors, columns, and pediments can be analyzed either individually or in conjunction with each other. In support of my hypothesis, I will quote here Palladio's own concept of beauty, which states, *La bellezza risulterà dalla bella forma, e dalla corrispondenza del tutto alle parti, delle parti fra loro, e di quelle al tutto: conciosiache gli edificij habbiano da parere uno intiero, e ben finito corpo...*[1750: I, 6]. In my opinion, the meaning of "correspondence" goes beyond the words, especially when we try to visualize it geometrically in a real building.

... On the windows

Just as a horizontal plane is suitable for establishing the width and length of a room, a vertical plane can be used to set the width and height of all the vertical elements. In practice, Palladio used both methods, as we will see. Despite the fact that the windows of the main rooms appear to be in the center of the walls in his elevation drawing – between the corner of the building and the interior face of the loggia's lateral arch –, they were actually built differently. Viewing the villa from the west (see fig. 4), it is evident that the window of the main room (*a*) is not placed at the center of the wall due to the fact of being next to the fireplace. In addition, this room has two more windows, one facing the interior of the loggia (*b*), and the other one around the building's corner (*c*). Therefore, the question to elucidate is how these windows were proportioned on the walls during construction. By tracing the diagonals of the wall (on fig. 4), it is remarkable that they intersect at exactly the left frame and upper corner of the window (*a*), and that this intersection point coincides with half of the height of the building. The same is true when the diagonals extend from top to bottom on the wall. Notice that the diagonals in both cases are taken from the interior face of the lateral arch of the loggia.

Fig. 4. A view from the northwest façade showing the windows' proportions as deduced in perspective. Photograph and superimposed drawing by the author

It appears more likely that the width of windows (*a*) and (*b*), and thus the fireplace in between of them, was proportioned along the large wall of the main room from its interior side; while the wrought-iron lattice of the window (*b*) run parallel to its diagonals.[2] In its turn, window (*c*) was put at the middle of the shortest wall in the same manner. This procedure may have been foreseen, since the walls of the *piano terra* were raised. In contrast, the heights of the windows seem to have been proportioned from the exterior, otherwise they would align with the lintel of the interior doors, below the halfway line of the building's height. The window sill also aligns with the lateral arch seat, as the grid makes evident. Once again, in fig. 4, if window (*a*) were placed above the crossing diagonals of this wall, then the human scale would be compromised, suggesting that only giant people could live there, and on the other hand, if it were placed below this diagonal, then the *piano nobile* would appear higher, suggesting that a window is missing, or that an entire floor is hidden. It would have simply been disastrous to give 1/8 of the wall's height above, or below its diagonals to set the window's height. Therefore, the intersection point of the diagonals is the one that strikes the perfect balance between the window and the wall, or *delle parti fra loro*, in Palladio's terms.

... On the lateral arches of the loggia

A rectangular grid successively divided through its diagonals and superimposed on one of the the lateral arches of the loggia reveals the geometrical construction of all of the arch's elements (see fig. 5).

Thus, the opening of the arch and its pilasters, the abutment and imposts, and the lower arch at the *piano terra* can all be determined. Of course, an exact match would be exceptional since measures ultimately depend on the building materials; for instance, the section of the pilasters depends on the size of the bricks. There are two rectangles in fig. 5, one from the top of the arch to the floor level, and the other from the top to the ground level. If the first one is divided into thirds vertically, its upper third would align at the spring of the imposts.

Obviously, such a division into thirds would coincide all along the shaft of the column, its section being one-ninth of the height of this rectangle, meaning that the proportion of the columns is 1:9. The diagonals of the second rectangle intersect at the spring line of the lower arch, fairly close to the pilaster section at one-fourth of its width. In the same manner, several other relations can be found in fig. 5 (facing page). What this grid is telling us is how the parts correspond to each other geometrically, and how the proportions were obtained by subdividing the whole into parts, as Palladio probably did in accordance with Vitruvius's definition of symmetry: *La symmetria è l'armonico accordo tra le parti di una stessa opera e la ripondenza dei singoli elementi all'imagine d'insieme della figura* [Vitruvius 1990: 23].

... On the loggias

The loggias, as projecting elements of the main body, are fundamental for the volumetric composition of the villa; without them the villa would be a simple hollow box. In addition, the stairs enhance the perspective of the visitor, leading him to a transitional space between outdoors and indoors. Thus, stairs, loggias, and main body were successfully combined in perspective as never before. Just like a pawn in a chess game has an important role to play, sometimes making the difference between winning or losing, the *piano terra* plays an important role in the Rotonda.

Fig. 5. A view of the loggia's arch from the northwest. Here, both the loggia's lateral size of 12 *f* and the columns' section of 2 *f*, were corroborated. The grid outlining makes the proportions among the parts comprehensible. Photograph and superimposed drawing by the author

If Palladio had not conceived a new architectural program for the villas, exploiting the the *piano terra* to free the *piano nobile* from the unsightly service spaces, and to avoid humidity as well, then you would not be reading this paper. The *piano terra* was precisely the key to raising the villa from the ground, making it appear more graceful from afar, while permitting a magnificent view of the landscape from the loggias.[3]

The incremented width of the central intercolumniation of the loggias makes the building appear more dynamic when one walks toward the stairs, while framing the access door at the same time; the lateral windows of the *piano nobile* appear centered in their walls (see fig. 6).

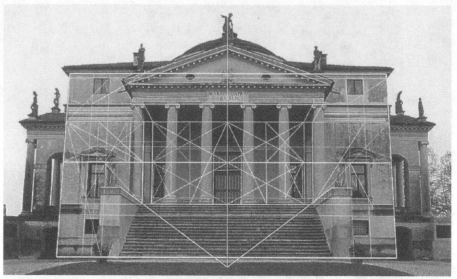

Fig. 6. The incremented breadth of the central intercolumniation of the loggia makes the building appear dynamic when walking towards the stairways, framing the access door while the lateral windows of the *piano nobile* appear to be centered in their respective walls. Photograph and superimposed drawing by the author

It might be naïve to ask what makes the loggias so magnificent, when we know that beauty has always been a controversial subject and depends on no specific formula. However, when my assistant Ambar Hernández and I were analyzing the northwest loggia, we found something that might help others to understand its beauty objectively, without trying to discover its formula. Common sense suggests that the columns must have been set along its transversal axis, which means that we cannot perceive their real size because our sight passes tangentially to both sides of the column shaft, instead of passing at the end points of their diameters.

In other words, a column in perspective looks a little bit wider than it actually is; this effect is, of course, more noticeable at close range. Therefore, we decided first to lay out a grid along the axis of the columns according to its intercolumniation sequence, and second, to lay out another grid tangential to the faces of the columns in order to investigate which type of sequence it could fit visually.

Thus, to establish the sequence of the former, according to Palladio's book, and our fig. 5, we have: 9d (18 f) for the height of the column, 6d (12 f) for the depth of the loggia, and 2 f for the columns diameter.[4] Therefore, the intercolumniation sequence along its axis is as follows:

$$4 \tfrac{1}{4} + 2 + 4 \tfrac{1}{4} + 2 + 5 + 2 + 4 \tfrac{1}{4} + 2 + 4 \tfrac{1}{4} = 30 f.$$

Now as a rule, to lay out the grid it has to be equally divided into as many parts as the sequence has feet. Therefore, the northwest loggia was divided in 30 f horizontally, and in 9d (18 f) vertically, as it is shown fig. 7a. Here, some vertical lines of the grid do not align with the shaft of the columns as expected, due to the sequence containing fourths of foot.

Fig. 7a. Here, a grid equally divided vertically in 30 f representing the colonnade sequence (4 ¼ + 2 + 4 ¼ + 2 + 5 + 2 + 4 ¼ + 2 + 4 ¼ = 30 f) along its transversal axis; does not visually fit with the columns' intercolumniation

Fig. 7b. Now, a grid equally divided vertically in 34 f' representing the colonnade sequence (5 + 2 + 5 + 2 + 6 + 2 + 5 + 2 + 5 = 34 f') up to its frontal plane, almost fits perfectly with the columns' intercolumniation

Fig. 7c. This new grid, placed at the wall of the piano nobile, also approximates the positions of the windows and the columnss. The conjunction of both virtual planes, 7b and 7c, produces the spatial consistency of the loggia. In 7c, the number of feet totals 66, which allows us to estimate the interior walls' thickness as 1 ¼ f, and the exterior's as 1 ½ f. Photograph by the author, with superimposed drawings by Ambar Hernández

As we already have pointed out, the viewer does not actually see the extreme ends of the columns' diameters, which is why we set the second grid tangentially to their faces. Thus, the viewer's sight would be captured when passing at both sides of the columns' entasis. So the question is, could such a grid coincide with the colonnade sequence? The answer is yes. According to the precepts of eustyle and diastyle, the intercolumniation is 2½ diameters (d) of a column for the former, and 3d for the second. These precepts were well known to Palladio from Vitruvius; as Palladio says, ... *mi proposi per maestro, e guida Vitruvio...* [1750: I, 5]. In geometrical terms an intercolumniation like this does not differ too much from the actual one, so we decided to apply it to the second grid and see what happened; unexpectedly it fit pretty well with the colonnade intervals, this is exactly the 'something we found' (see fig. 7b).

The roundness of the columns makes the tangential plane more suitable because it corrects the visual appreciation of the geometrical sequence in perspective. As can be appreciated in fig. 7b, it is remarkable how the eustyle-diastyle grid matches with both empty and solid spaces along the façade of the loggia. Therefore, the grid sequence:

$$5 + 2 + 5 + 2 + 6 + 2 + 5 + 2 + 5 = 34 f',$$

is an interpretation of the proportions of the loggia as they are perceived in perspective. Naturally, f' measures a little bit less than f. This discovery leads me to hypothesize that Palladio was aware of these sorts of perspective effects, and corrected the Vituvian formula as it should be when applied in perspective. It seems paradoxical that $30 \times 18\ f$ can transform into $34 \times 18\ f'$ while maintaining a ratio of 1.667.

Finally, by laying out a new grid on the building's façade from side to side, similar to that of fig. 7b, but now in true Vicentine feet, it once again makes sense with the eustyle-diastyle sequence, as is shown in fig. 7c. Even though this grid does not fit all the elements in the sequence perfectly (and I did not want to force it), it suggests that it could fit perfectly at certain intervals of distance. As I have learned from my practice in perspective, distance is always involved. Proving this would require more photographs taken from a distance one meter closer or further away. However, both the grids in figs. 7b and 7c fairly depict the spatial consistency of the depth of the loggia, since they echo each other visually. This explains why the loggia appears so well proportioned to the whole from any distance or when we move closer to it, and perhaps it is this invisible echo that results in its beauty.

Conclusion

This is a glimpse of what can be achieved by superimposing grids on photographs in order to analyze the Rotonda's actual proportions. Here, I have only focused on some parts of the whole. A complete analysis of the windows with all their ornamental elements would itself require another paper, and so on for the rest of parts. The grid criteria could be applied in the same way to analyze the constructive system of the building, since walls and vaults were also built in proportion. In particular, the brick vaults of the *piano terra*, which I was lucky enough to capture with my own camera in 2003 (see fig. 8), cry out for us to elucidate their constructive geometry. After all, the *piano terra* was the key to the innovation embodied by the Rotonda, as well as one key reason why this building was included in the list of Unesco World Heritage Sites.

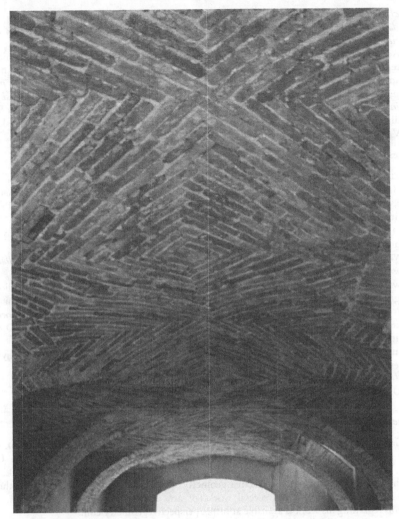

Fig. 8. This photo of the southwest passage at the piano terra shows us the
constructive system of the vaults with its peculiar interlaced brick pattern. Their
proportions and geometry will be the subject of future study. Photograph by the
author

Notes

1. *Onde perche gode da ogni parte di bellissime viste, delle quali alcune sono terminate, alcune
   più lontane, & altre, che terminano con l'Orizonte; vi sono state fatte le loggie in tutte
   quattro le faccie...*[Palladio 1570: II, 18].
2. In my opinion, the window-wrought-iron of window (b) was deduced from the proportion of
   the opening, not by following a predetermined angle.
3. *Però lodo che nella più bassa parte della fabrica, laquale io faccio alquanto sotterra; siano
   disposte le cantine, i magazini da legne, le dispense, le cucine, i tinelli, i luoghi da liscia, o
   bucata, i forni, e gli altri simili, che all'uso quotidiano sono neccessarij: dal che si cavano due
   commodità: l'una che la parte di sopra resta tutta libera, e l'altra, che non meno importa; è,*

che detto ordine di sopra diviensano per habitarvi, essendo il suo pavimento lontano dall'humido della terra: oltra che alzandosi; ha più bella gratia ad esser veduto, & al veder fuori [Palladio 1750: II, 3- 4].

4. According to Palladio [1750: II, 4] and [1750: III, 6], it is supposed that the graphical scale representing half a Vicentine foot is drawn at full scale, in which case it measures approximately 35 cm. The note below the graphical scale says: *Questa linea è la metà del piede Vicentino, co'l quale sono state misurate le seguenti fabrique. Tutto il piede si partisce in oncie dodici, e ciascun'oncia in quattro minuti.* According to Semenzato [1990: 92], the diameter of the columns measured at the base is equal to 0.745 m, supposedly equivalent to 2 Vicentine feet, in which case $f = 0.372$ m.

References

BERTOTTI-SCAMOZZI, Ottavio. 1786. *Le fabbriche e i disegni di Andrea Palladio.* Venice.

GARCÍA-SALGADO, Tomás. 2004. La Invención de la Rotonda. http://www.perspectivegeometry.com

GIACONI, Giovanni and Kim WILLIAMS. 2003. *The Villas of Palladio.* New York: Princeton Architectural Press.

PALLADIO, Andrea. 1570. *I quattro libri dell'architettura.* Venice. Facsimile edition, Ulrico Hoepli Editore Libraio, Milan, 1990.

SEMENZATO, Camillo. 1990. L'architettura della Rotonda. Pp. 45-100 in *Andrea Palladio. La Rotonda*, Renato Cevese, Paola Marini and Maria Vittoria Pellizzari, eds. Milan: Electa.

SASS, Lawrence. 2001. Reconstructing Palladio's Villas: A Computational Analysis of Palladio's Villa Design and Construction Process. Pp. 212-226 in *Reinventing the Discourse – How Digital Tools Help Bridge and Transform Research, Education and Practice in Architecture* [Proceedings of the Twenty First Annual Conference of the Association for Computer-Aided Design in Architecture (ACADIA), Buffalo NY, 11-14 October 2001.

———. n.d. Materializing Palladio's Ideal Villas. Computational Reconstruction through Physical Representation of Digital Information.
http://ddf.mit.edu/projects/PALLADIO/White_Paper_a_Palladio.pdf.

VITRUVIUS, Marco Pollione. 1990. *De Architectura, Libri X.* Padua: Edizioni Studio Tesi.

About the author

Tomás García-Salgado received his professional degree (1968), Master's degree and Ph.D. (1982-1984) in architecture. He is a formal researcher in the Faculty of Architecture of the UNAM (México), and holds the distinction as National Researcher, at level III. Since the late 1960s, he has devoted his time to research in perspective geometry, his main achievement being the theory of Modular Perspective. He also has several works of art, architecture, and urban design. More information regarding his work is available at http://perspectivegeometry.com.

Carl Bovill

School of Architecture,
Planning, and Preservation
University of Maryland
College Park, MD 20742
bovill@umd.edu

Keywords: fractals, geometry,
Doric Order, iterated function
system, Owen Jones

Research

The Doric Order as a Fractal

Abstract. Owen Jones in *The Grammar of Ornament* clearly states that ornament comes from a deep observation of nature. He emphasizes the importance of the harmony of the parts and the subordination of one part to another. This subordination and harmony between the parts is what fractal geometry explores as self-similarity and self-affinity. An iterated function system (IFS) is a digital method of producing fractals. An IFS in the shape of columns holding up a lintel produced an attractor displaying fluted columns with capitals and an entablature with the proper number and spacing of triglyphs and mutules. Thus, fractal geometry, through the use of iterated function systems, provides a new insight into the intention of Doric ornament design.

Fractal geometry

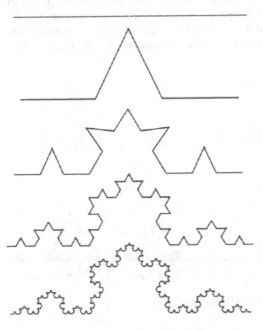

Fig. 1. Koch curve [from Bovill 1996]

Fractal geometry is the study of self-similar geometric forms. An easy to understand fractal is the Koch curve shown in fig. 1. The Koch curve was first drawn in 1904 by mathematician Helge von Koch. He drew it as a question to mathematicians about continuous functions that were not smooth and thus did not have a tangent [Mandelbrot 1983: 35]. The Koch curve is constructed in stages. The first stage is constructed by dividing a straight line into three equal parts. The middle part is taken out and replaced with an equilateral triangle with the base removed. In the second stage the same process described above is repeated on each of the four straight line segments produced by the first stage. This process is continued to infinity. This procedure, where the output from the previous stage is used as the input for the next stage, is called iteration in mathematics.

Iteration is at the heart of current mathematical research into nonlinear dynamics and chaos. One can zoom in on a portion of the Koch curve and observe detail similar to the entire curve. The smaller sections of the curve are self-similar to larger parts of the curve. Self-similar in mathematics means that a reduction, enlargement, and or translation has been made without distorting the relative size of lengths and angles [Peitgen Jurgens Saup 1992: 138-146].

The Koch curve can also be constructed with a random component to it. A random Koch curve is constructed in the same way as the regular Koch curve except that the up or down orientation of the equilateral triangle is determined by a flip of a coin. The smaller parts of the random Koch curve are self-affine to the larger parts. They are similar but not identical. In mathematics a self-affine transformation can distort relative lengths and angles or create mirror images as it reduces, enlarges, and or translates the original [Peitgen Jurgens Saup 1992: 138-146].

A tree is a natural example of a fractal. The branching is self-affine through three to four levels of branching. The self-affine branching creates the harmonious relation between the main branching and the second, third, and fourth levels of branching.

Iterated function systems

One method of producing fractal shapes is with an iterated function system (IFS). In an IFS, an image is reduced and placed on top of itself over and over again. Peitgen, Jurgens and Saup use an analogy of a copy machine with multiple reducing lenses [1992: 23-36]. The Sierpinski gasket can be made in this fashion with three, one-half reductions placed in the form of an equilateral triangle. The nature of the transformations determines the appearance of the final image. The shape of the original image has no effect on the outcome [Peitgen Jurgens Saup 1992: 24]. Fig. 2 shows this process starting with a square and with a circle.

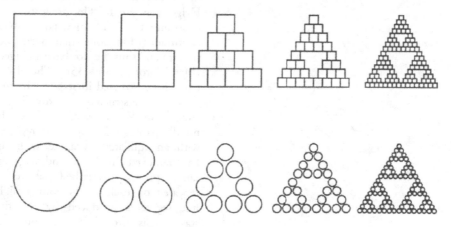

Fig. 2. Sierpinski gasket [from Bovill 1996]

Fig. 3 shows an IFS that produces a Koch curve. The resulting image is referred to as the attractor. The attractors shown in fig. 3 were drawn by hand using tracing paper overlays to produce each succeeding stage of the fractal construction. This hand-produced iteration is helpful in understanding what an IFS does. However, exploring more complicated IFS transformations requires a computer program [Bovill 1996: 47-53].

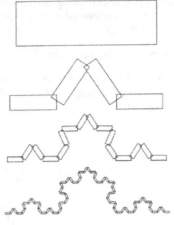

Fig. 3. Koch curve produced as an iterated function system [from Bovill 1996]

A computer IFS was used as a research tool to explore where self-similarity or self-affinity might lurk in Doric temple ornament. Michael Barnsley's *Desktop Fractal Design System* [1992] was the IFS that was used. In this system rectangles of any size and orientation are placed on a background square. Each rectangle represents a self-affine transformation. The system uses a random process to determine a set of points that will be on the final attractor. As the program runs, more and more points are determined, displaying an ever improving image of the final attractor [Peitgen Jurgens Saup 1992: 297-306].

The Grammar of Ornament

In *The Grammar of Ornament* Owen Jones clearly states that ornament comes out of an observation of nature's deep structure [1972: 22-25]. In the following quote note the statements about the harmony of the parts and the subordination of one part to another.

> As we proceed with other styles, we shall see that they approach perfection only so far as they follow, in common with the Egyptians, the true principles to be observed in every flower that grows. Like these favorites of nature, every ornament should have its perfume: i.e. the reason of its application. It should endeavor to rival the grace of construction, the harmony of its varied forms, and due proportion and subordination of one part to the other found in the model. When we find any of these characteristics wanting in a work of ornament, we may be sure that it belongs to a borrowed style, where the spirit which animated the original work has been lost in the copy [Jones 1972: 24].

In his discussion of Greek ornament Jones states, "What is evident is, that the Greeks in their ornament were close observers of nature, and although they did not copy, or attempt to imitate, they worked on the same principles" [1972: 33]. A close observation of nature, aided by fractal awareness, reveals that self-similarity and self-affinity provides the harmony and subordination of one part to another. Jones continues on with the observation that one of the great features of Greek ornament is its application of these learned principles of nature in the most humble as well as the highest works of art [Jones 1972: 31-37]. This is paralleled in nature. Nature displays a fractal complexity from the organization of galaxies in space to the structure of gold clusters observed with a transmission electron microscope [Mandelbrot 1983: 84-96; Feder 1988: 38].

Iteration of a group of columns holding up a lintel

If a set of rectangles defining an IFS is arranged as a sequence of columns holding up a lintel, the resulting attractor is a group of fluted columns with a cascade of ever shorter fluted columns on top of them (fig. 4).

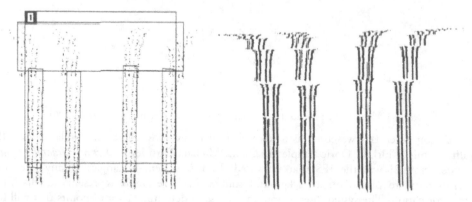

Fig. 4. Iterating columns holding up a lintel produces fluted columns

This suggests that fluting of columns is more than a holdover from the forming of wood columns with a curved adze, or an emphasis of the vertical nature of the column, as suggested by Elisabeth Ayrton in *The Doric Temple* [1961: 151]. Fluted columns with capitals is the attractor, or the deep image, of a group of columns supporting a lintel. Stated another way, each fluted column and capital is a self-affine contraction of the whole temple. This can be achieved by a transformation that takes the whole temple and contracts it horizontally to the width of a column. The Greeks were expert in geometry and in stone working. If a deep image of fluted columns did not exist they would probably have tended to a more perfect circular shape.

Doric temple columns had from twelve to twenty-four flutes, with the most common number of flutes being twenty. As pointed out by James Curl in *Classical Architecture*, twenty flutes was a good number because it results in an arris (the edge formed where two flutes meet) under the corner of the square abacus and a flute in the center of the front of the column [Curl 1992: 22]. This number of flutes is interesting since it produces an image on the front of the column of eight arrises which is the same as the eight columns that are across the front of the Parthenon, the ultimate Greek temple. The IFS produces a number of arrises equal to the number of columns in the original set up of the IFS. In fig. 4 there are four arrises because the IFS had four columns.

Another interesting issue involves the column spacing at the temple corners. Could the slightly smaller spacing between columns at the temple corners be partially a reflection of the fact that the arrises near the edge of a column appear closer together when observed from a distance. Is it possible that the variation in column spacing is a self-affine echo of the elevation view of a fluted column? In fig. 4, the columns were spaced closer together at the edges. The center spacing is larger. The resulting attractor has arrises that are spaced closer together at the edges. In addition consider how the columns were constructed. In *The Greek Temple*, Isabel Grinnell points out that the columns were made of several drums fastened together. The lowest drum and the highest drum including the necking of the capital were fluted before being placed. The other drums were fluted after being placed [Grinnell 1943: xviii]. Now think about the temple designers looking at the upper drum. The upper drum is clearly a little model of the entire temple with the arrises representing columns and the capital the entablature. Some Ionic temples actually had column spacing across the front of the temple that got progressively closer together from the center bay out. A. W. Lawrence, in *Greek Architecture*, describes the Ionic temple the Heracum at Samos. He points out that the column spacing was graduated to emphasis the center entrance [Lawrence 1983: 162].

Iteration of a group of columns with capitals holding up a lintel

To explore this attractor further, simple capitals were added to the columns supporting the lintel. This new IFS was then run to produce the attractor shown as fig. 5.

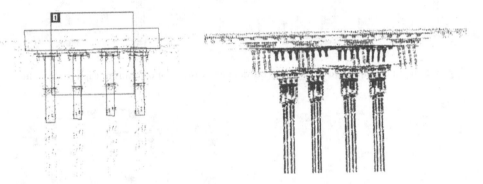

Fig. 5. Iteration of columns with capitals holding up a lintel

The fluted columns with capitals are still there but another feature has appeared. There are multiple echoes of the fluting of the columns hovering above the tops of the columns in the location where the triglyphs and mutules with their guttae would be in a Doric Temple. Are the triglyphs – and then, at a smaller scale, the mutules and guttae – fractal echoes of the entire temple? Also note that one can interpret fig. 5 in two ways. It can be seen as a group of four fluted columns with capitals holding up an entablature, or it can be seen as one large fluted column with a complex capital. The whole temple and the fluted column are echoes of each other.

The doubling cascade of Doric temples

The images produced with the IFS strongly suggest that there is considerable self-affinity in the ornamental detailing of Greek temples. This conclusion is not in conflict with any of the historically stated reasons for fluting or triglyphs; it just overlays a strong reason for their continued use as ornamental features on stone temples. Fig. 6 is an iteration of a group of columns holding up a lintel where the lintel is made up from two pieces. The two-piece lintel sets up a doubling cascade. In a Doric Temple there are one less than two triglyphs for each column (for example, for eight columns there are seven triglyphs), and there are one less than two mutules for each triglyph. The slight overlap of the two piece lintel in the center takes care of the "one less than" situation. This iteration produces fluted columns with capitals and an entablature with the proper number and spacing of triglyphs and mutules. Fig. 7 adds a pediment to the iteration. This causes more complex column capitals (Ionic, Corinthian) and lots of dentils.

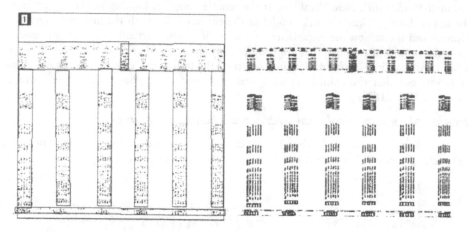

Fig. 6. Making the lintel two parts sets up the doubling cascade of columns to triglyghs to mutules

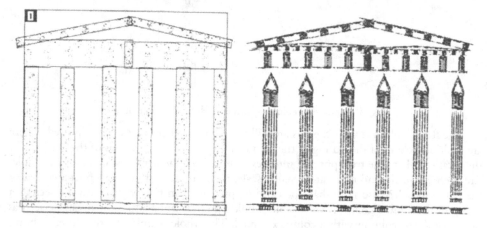

Fig. 7. A pediment added to the iteration

Conclusion

The IFS analysis of a Doric temple clearly shows a self-affine cascade of ornament that is four levels deep: the entire temple, the column and capital, the triglyphs, and finally the mutules and guttae. A tree or leaf has this same order of magnitude of self-affine branching. Is this coincidence, or were the Greeks more observant than we think? Remember Owen Jones's statement about Greek designs in *The Grammar of Ornament*, "What is evident is, that the Greeks in their ornament were close observers of nature, and although they did not copy, or attempt to imitate, they worked on the same principles" [1972: 33].

References

AYRTON, Elisabeth. 1961. *The Doric Temple*. New York: Clarkson N. Potter.

BARNSLEY, Michael. 1992. *The Desktop Fractal design System*. Boston: Academic Press.

BOVILL, Carl. 1996. *Fractal Geometry in Architecture and Design*. Boston: Birkhäuser.

CURL, James. 1992. *Classical Architecture*. London: Batsford Ltd.

FEDER, Jens. 1988. *Fractals*. New York: Plenum Press.

GRINNELL, Isabel. 1943. *Greek Temples*. New York: The Metropolitan Museum of Art.

JONES, Owen. 1972. *The Grammar of Ornament* (1st ed. 1856). New York: Van Nostrand Reinhold.

LAWRENCE, A.W. 1983. *Greek Architecture*. New Haven: Yale University Press.

MANDELBROT, Benoit. 1983. *The Fractal Geometry of Nature*. New York: W.H. Freeman.

PEITGEN, Heinz-Otto, Hartmut JURGENS and Dietmar SAUP. 1992. *Chaos and Fractals*. New York: Springer.

About the author

Carl Bovill has a bachelor's and master's degree in mechanical engineering form the University of California, and a master's degree in architecture from the University of Hawaii where he earned the American Institute of Architects Gold Medal for excellence in the study of architecture. Professor Bovill teaches materials and methods of construction, environmental control systems, and technical system integration in design studio and, in collaboration with other faculty, received an American Institute of Architects Education Honors Award in 1995. He has also taught at California Polytechnic State University in San Luis Obispo, and at the University of Tennessee in Knoxville. Prior to teaching Professor Bovill was a member of Interactive Resources, Inc., an architectural firm in Richmond, California. His publications include *Architectural Design, Integration of Structural and Environmental Control Systems*, Van Nostrand Reinhold, 1991, and *Fractal Geometry in Architecture and Design*, Birkhäuser, 1996. Some of the ideas in this paper were first presented at the Millennial Open Symposium on the Arts and International Computing Conference held at the University of Washington, Seattle, Washington in August 2000.

Michael C. Duddy

MDA Designgroup
International
19 Union Square West
New York, NY 10003 USA
MDuddy@mda-
designgroup.com

Keywords: Doric temple, Greek
architecture, Parthenon, optics,
perspective, Euclidean
geometry, non-Euclidean
geometry, projective geometry,
hyperbolic visual space

Research

Roaming Point Perspective: A Dynamic Interpretation of the Visual Refinements of the Greek Doric Temple

Abstract. Writers, artists, and mathematicians since Vitruvius have attributed the use of the visual refinements as a means by which the Greek builders optically corrected the form of the Doric temple. This study proposes an interpretation in which the visual refinements of the Parthenon are considered from a non-stationary, or "roaming", point of view. The mathematics of this type of visual space reveals a dynamic zone in which objects visually increase and decrease simultaneously, a behavior consistent with conditions addressed by the visual refinements of the Parthenon.

*The Greeks created a plastic system by forcibly affecting our senses. ....*They employed the most delicate distortions, applying to their contours an impeccable adjustment to the laws of optics.

Le Corbusier, *Towards a New Architecture*

Introduction

In the outline of his program for the classical orders in the *Ordonnance for the Five Kinds of Columns after the Method of the Ancients*, Claude Perrault attacks what he claims are the "abuses" of the orders, specifically the use of optical corrections or visual refinements. From the earliest Doric temples of ancient Greece, angular and curvilinear irregularities could be observed in the building forms that included the tapering of the columns, curvatures in the podiums and entablatures, inclinations of columns and friezes, and variations in column diameters and spacings. For Perrault, these adjustments opposed an otherwise rigorous system of logically proportioned elements, "distorting or spoiling proportions in order to prevent them from appearing distorted or for making something defective in order to correct it" [Perrault 1993: 162-163]. Perrault's indictment references back to Vitruvius who, in setting out in Book III of the *Ten Books of Architecture* a detailed proportional system of the Classical orders, mentions that despite his exacting prescription, adjustments to the proportions of columns were necessary to compensate for optical distortions. "For the eye is always in search of beauty, and if we do not gratify its desire for pleasure by a proportionate enlargement in these measures, and thus make compensation for ocular deception, a clumsy and awkward appearance will be presented to the beholder" [Vitruvius 1960: 86]. Not only are the columns adjusted to account for such distortions, but "the level of the stylobate must be increased along the middle by the *scamilli impares*; for if it is laid out perfectly level, it would look to the eye as though were hollowed a little."[1] In these recommendations, Vitruvius is accounting for the Greek and Roman convention of using the curved geometries in the columns, podiums, and entablatures was to "correct" optical distortions. Subsequent architects from Alberti to Le

Corbusier would note how the use of these "visual refinements" offered evidence of a higher order of visual understanding.[2]

This study began as an inquiry into the so-called "subjective" or curved model of visual space as a possible explanation for the visual refinements. Armed with an understanding of the non-Euclidean geometries of Bolyai and Lobachevski, writers from Hermann von Helmholtz in 1856[3] to Rudolph Luneburg in 1947[4] claimed that the geometry of visual experience actually behaved according to a hyperbolic construct. To substantiate their claims they would cite the subtle curvatures of the Greek visual refinements as evidence, noting that the visual refinements somehow acted as the inverse complement of the shape of visual space counteracting its curved geometry (fig. 1).

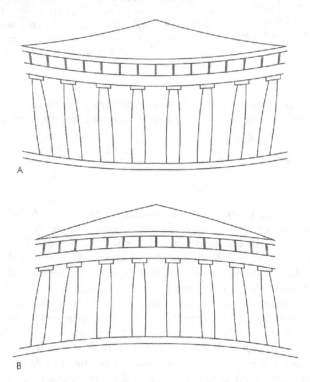

Fig. 1. When considered together the optical refinements of temple B would counter the apparent visual form of temple A to yield a temple in visual equilibrium.
Drawing by the author after [Fletcher 1967: 95]

In the course of my research several points would direct my investigation: (1) The proponents of curved visual space never demonstrate how the refinements actually corroborate their model of curved visual space; (2) The refinements do not correspond to a concise mathematical model left to us by the Greeks; (3) The temple was never meant to be perceived as a static object from a fixed point of view. My premise, by extension, asks whether the refinements could qualify a different type of visual experience; that of a dynamic or roaming point of view. Perhaps, as opposed to the conventional linear and curved methodologies taken from a stationary view point where a three-dimensional form is projected onto a two-dimensional surface, the visual refinements might account for an alternative condition in which a four-dimensional space-time is projected onto a three-dimensional form. What follows is a mathematically-based proposal for a dynamical consideration of the visual refinements as applied to the Parthenon, considered by most scholars as the exemplary Greek Doric temple.[5]

From a Fixed Point of View

Following Vitruvius, it is generally assumed that the visual refinements were employed to compensate for some type of optical effect present in the temple.[6] "Dipping", "bowing", "sagging" are some of the ways this effect has been characterized, and explanations of optical effects typically rely on visual or psychological analyses derived from interpretations of subjectively-experienced phenomena. For example, the fact that straight parallel lines might appear bent when viewed in the context of adjacent skewed or curved lines provides a possible explanation as to why entasis was introduced. Similarly, if light-colored background fields can make foreground figures seem smaller, increasing the diameters of the corner columns should therefore offset their diminished appearance "because they are sharply outlined by unobstructed air round them, and seem to the beholder more slender than they are" [Vitruvius 1960: 84]. One approach to representing an aspect of the subjective condition is by linearly projecting an image onto a picture plane from the stationary view point of the observer. Linear perspective, as it is known, would eventually be generalized by Desargues to form the foundation of projective geometry.[7]

When the Parthenon is considered from a subjective point of view, we can first ask whether it behaves according to the logic of linear perspective as Perrault assumed.[8] Had the Greeks intended that the Parthenon be perceived from a single position or from several specific positions according to the principles of linear perspective, it would follow that a fixed point or multiple points can be located from where the visual refinements, when considered collectively, would allow the temple to appear "correct". These visual snapshots – singular views from multiple fixed points – are predicated on establishing specified vantage points from which visual space would be linearly constructed on essentially a two-dimensional pictorial surface in space. One can easily experience the effects of perspectival diminution, for example, by viewing Andrea Pozzo's famous ceiling and "domed" crossing in Sant Ignazio from any point other than the center of the nave. Similarly, if in three-dimensional physical space, an optically adjusted form constructed from a single point only appears correct from a single station point, then that form would seemingly appear distorted from all other points. In the case of Bernini's windows of the corridor leading to the Scala Regia that flank the piazzeta of San Pietro, the openings will appear correct from the ellipse of the Piazza San Pietro but are clearly "distorted" rhombic shapes when viewed from the piazzeta. Hence, following this reasoning, where might these points be located on the Acropolis where the Parthenon appears "correct"?

As with most of the acropoli of ancient Greece, the approach to the Parthenon is a sophisticated and elaborately choreographed procession (fig. 2).

Emerging from the east peristyle of the Propylaea, the processioner encounters the classic three-quarters view of the Parthenon sitting on the rise off to the right where both the north and western peristyles are in equal view. From here the path to the naos of Athena proceeds along the north side of the temple, bypassing completely its eastern front to a point near the northeast corner where it turns right toward the flight of stairs leading to the platform on which the stereobate sits. At this point, the temple can again be perceived from a three-quarters view albeit from close proximity. From here the processioner continues to a position directly to the center of the east octastyle where upon turning right a frontal view of the east elevation is obtained at close range; the steps in the stereobate leading to the naos of Athena lies directly ahead.

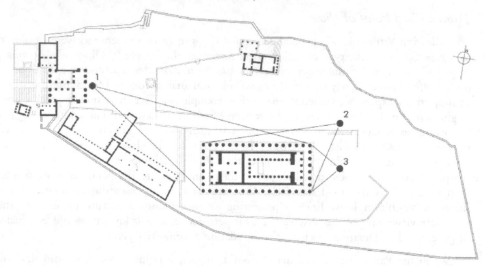

Fig. 2. Plan of the Acropolis showing the processional pathway from Propylaea to naos. Drawing by the author after the plan of the acropolis is [Scully 1991]

Along this route we can identify three specific points of reorientation where the Greek builders might have intended for the traveler to pause and visually evaluate the Parthenon before proceeding. This suggests, of course, that these might be the likely points where it is essential that the temple appear "correct" and where the visual refinements fulfill their purpose. Nevertheless, can the Parthenon appear "correct" from all three points simultaneously? Can a three-quarters view appear equally correct from close proximity and from a point six times that distance? Can the frontal view from close proximity appear "correct" with the same adjustments that make the distant three-quarter view appear correct? According to the method of linear perspective, one station point necessarily has to prevail over the others; there can be only one point where the temple will appear "correct". Perhaps the Greek builders intended that each individual refinement be specific to one position and not another, and that the cumulative effect of the visual refinements would be no more than an average of an optically adjusted temple where the correct and the non-correct interchange depending on position. Or, possibly, the Greeks may have devised an alternate method that unified the visual refinements, a construction that is not linear at all, but curvilinear.

Curved Considerations

While linear perspectival construction offers a reliable method to represent a three-dimensional space on a two-dimensional surface, it nevertheless proves to be inconsistent with the geometry of perceived reality, for even at the time when the methods of linear perspective were being perfected, artists were already confounded by the visual distortions within the peripheral fields of their works, especially when projecting objects within a plane onto a parallel picture plane. Leonardo demonstrated in the *Codex Atlanticus* how these geometrical inconsistencies could be demonstrated by a row of equally sized and equally spaced columns projected onto a parallel plane from a central station point; the distant columns are rendered larger than those in the foreground (fig. 3).

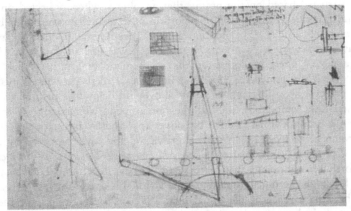

Fig. 3. Leonardo da Vinci, detail of a sketch from *Codex Atlanticus*, fol. 520 r showing the inconsistencies of linear perspective when a row of equally spaced columns are projected onto a parallel plane. The distant columns will appear larger than those in the foreground

The geometry of parallel plane projection could not reconcile these types of marginal distortions and brought into question whether distances projected onto the picture plane were isomorphic with the three-dimensional form.[9]   In response, some commentators considered instead the angular method where size and scale are determined by angular increments along a circular arc centered at the station point rather than relating lengths on a picture plane using the distance method of linear perspective (fig. 4).

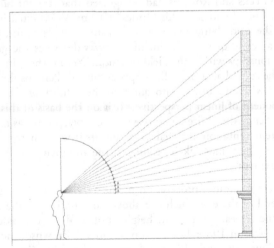

Fig. 4. Serlio's diagram of angle measure demonstrates that equal view angles correspond to increased intervals in a parallel plane projection. Drawing by the author based on [Serlio 1982]

On the basis of the angle method, subsequent writers were to argue that our visual field was in fact curvilinear, and that representing reality required acknowledging the subtle curves that describe our visual space where vertical and horizontal lines are represented as curves converging on distant points. Acknowledging the angle method proposed by Euclid in the *Optics*, these writers speculated that a rigorous system of curvilinear perspective could be constructed based on the notion that an object, when projected on the curved surface of our visual field, would occupy a portion of that surface determined by the angle limits as described by Euclid. The graphic approach, championed by Erwin Panofsky in his seminal essay, *Perspective as Symbolic Form* [1927], was first suggested by Herman von Helmholtz and was later developed by the mathematician Guido Hauck and others [Panofsky 1991: 27-36 and n. 12, 87-92]. In Hauck's method, where vanishing points were fixed on the horizon line and at points above and below the field of vision, vertical and horizontal lines would bow around the horizon line, converging on the vanishing points thereby creating a visual "shield" of convex space before the stationary station point of the observer. More recently, a rigorous analytic method was proposed by Rudolph K. Luneburg, who, in his investigation, *The Mathematical Analysis of Binocular Vision* [1947] set forth a methodology that mathematically described a non-Euclidean hyperbolic visual space based on the transformation of the visual shield as it moves between the near-space and the distant-space of the visual field. Nevertheless, in spite of these detailed graphic and analytic demonstrations, commentators continue to disagree as to whether the interpretation of the empirical results justifies the mathematical and graphic models proposed to represent curved visual space.[10]

While the science of curvilinear perspective is beyond the scope of this paper it should be emphasized that the proponents of curved visual space were consistent in citing the ancient Greek and Roman use of the visual refinements as evidence to support their claims. From Gombrich[11] to Heelan[12], writers contend that the ancient builders were the first to acknowledge the curvature of visual space, yet they never provide a demonstration nor offer a proof to support their claim. Kline attributes the curvilinear space directly to Euclid, claiming that "the Greeks and Romans had recognized that straight lines appeared curved to the eye. Indeed, Euclid said so in his Optics" ([Kline 1954] cited in [Brownson 1980: 181]). Based on the underlying premise that from a single point, visual rays can be projected to form a "cone of vision", Euclid's *Optics* describes the geometry of various conditions and relationships within this field of vision. Yet nowhere in the 58 propositions of the *Optics* does he state that straight lines appear curved. Perhaps the closest he comes is in his Proposition 8 where he brings into question the validity of the distance method of scaling, a key component of linear perspective. It is on the basis of this Proposition 8 that Panofsky would challenge the entire enterprise of linear perspective as nothing more than a convention for the representation of visual space, not its true geometry.[13] Since Kline never substantiates his claim, he is most likely relying on the arguments proffered by Panofsky to support his bold statement.

Interestingly, there is another proposition in the *Optics* that states that "in the case of objects below the level of the eye which rise above one another, as the eye approaches the objects, the taller one appears to gain in height, but as the eye recedes, the shorter one appears to gain" [Brownson 1980: 177]. Proposition 15, Brownson notes, along with the associated Propositions 16 and 17 stand as unrelated oddities among the other Propositions, as they are predicated on "the special arrangement of objects which are not mentioned in the statement" [Brownson 1980: 177]. However, let us consider a stack equally-sized objects one above the other in a position below the level of the eye. As we approach the stack we would anticipate that the ones at the top, closer to our eye, would increase to a greater extent than those lower in the stack because relatively the increase in their angular measure is greater. Similarly, when receding, we might anticipate that the inverse will occur; the lower objects would "appear to gain" since the angles of the upper objects are decreasing more rapidly. But in fact Euclid's proposition only holds in certain positions of the viewer, for with a constantly moving view point, or roaming point, the rate of angular change varies. In some positions within what I will call the *dynamic zone*, some view angles are increasing while others are decreasing.

Roaming Point Perspective

To get a sense of a roaming point of view, consider the phenomenal effect of approaching a large upright planar surface such as a uniform brick wall. As we advance forward, the wall begins to fill our cone of vision as the view angles of the individual bricks increase. At the point where we enter the dynamic zone the wall expands beyond the limits of our visual cone and we begin to notice that the peripheral edges of the wall seem to snap back at a rapid rate in the direction of our movement. As our eyes near the surface of the wall the rate of "snapping" accelerates until the periphery settles into the plane of the wall as our eyes become coplanar with it. Reversing the movement produces the opposite effect; as we back away the periphery snaps forward to embrace us and as we reach the distant zone where the wall is comfortably contained in our cone of vision, the periphery again settles into a flat plane. The impression is that of a "visual shield" similar to that described by Panofsky that pulses in and out depending on the direction of our movement.

Consider a line *m* on which three distinct points *O*, **a** and **b** are equidistantly spaced at a distance **D** from each other (fig. 5).

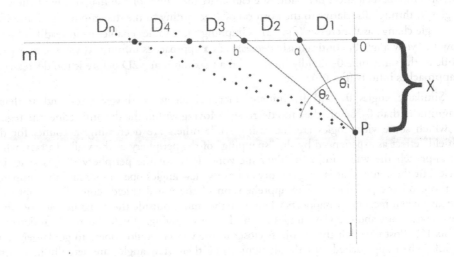

Fig. 5. As *P* moves away from *O* along *l*, θ_1 continuously decreases from 90°. At the same time, however, θ_2 increases until the point *P* reaches $x = \sqrt{2}D$ when it too decreases as *x* approaches infinity

Construct a line *l* perpendicular to *m* at *O*. Now, place a point *P* on *l* through which lines extended from **a** and **b** will pass. Clearly the angles θ_1 and θ_2 formed at *P* will vary in size, with θ_1 being the larger and θ_2 the smaller. We can also surmise that as *P* moves along *l*, both the angle measure of θ_1 and θ_2 will vary as well depending on the position of *P*. For example when *P* lies at *O*, $\theta_1 = 90°$ while $\theta_2 = 0°$. Similarly, as *P* approaches infinity on *l*, angles θ_1 and θ_2 will approach zero. Between the two limits let us allow *P* to "roam" along *l*, maintaining a distance *x* from the point *O*. As *P* moves away from *O* on *l*, θ_1 diminishes continuously from 90° to 0° as *x* approaches infinity. However, angle θ_2 increases as *x* increases near *O*, and reaches a maximum value before it too diminishes to 0 as *x* approaches infinity. Using right triangle trigonometry, we see that $\tan \theta_1 = D/x$, and for $n=1,2,3 \dots$,

$$\theta_n = \tan^{-1}\left(\frac{nD}{x}\right) - \tan^{-1}\left(\frac{(n-1)D}{x}\right). \tag{1}$$

Taking the derivative $d\theta_n/dx$ and setting it equal to zero, we find that the maximum value of θ_n, for each $n = 2, 3, \dots$, is

$$\theta_n^{\max} = \tan^{-1}\sqrt{\frac{n}{n-1}} - \tan^{-1}\sqrt{\frac{n-1}{n}}, \tag{2}$$

which occurs at the *x* value

$$x_n^{\max} = \sqrt{n(n-1)}D. \tag{3}$$

As a result of our roaming view point, the apparent shifting of the visual shield between convex and concave according to the "snapping" of the peripheral limits of the wall follows this specific mathematical behavior. We can chart the values of the magnitude of the view angles θ_1 through θ_4 relative to the position of *P* according to the diagram in fig. 5 and plot the angle change as *x* increases (fig. 6a). Graphing these values enables us to readily observe how the view angle θ_1 continuously decreases as *x* approaches infinity, as we would expect, while at the same time θ_2 rapidly increases to a maximum at $\sqrt{2}\mathbf{D}$ before it too decreases as x approaches infinity (fig. 6b).

Similarly angles θ_3 and θ_4 initially increase, albeit at slower rates and to lesser magnitudes than θ_2, before they too decrease. Moving within the dynamic zone, the region in which some view angles are increasing while others are decreasing, accounts for the "shield" effect as experienced by the "snapping" of the periphery of the wall. In fact, when we approach the wall within the dynamic zone, it is not the periphery that is snapping back, but the center that is surging forward as its view angles open more rapidly compared to those on the periphery. The apprehension of the visual shield, convex on approach, concave when receding, disappears when we either move outside the dynamic zone or when the viewer stops and the view angles are no longer changing. Euclid correctly understood in his 15th Postulate that those objects closer to the viewer would appear to get larger more rapidly when approached. In the dynamic zone their view angles are actually increasing while those of the more distant objects are diminishing. What Euclid does not account for, however, is the fact that within the near zone, the closer objects are actually visually *decreasing* at a greater rate than the distant objects, and in the distant zone all the objects appear to gain at a perceptually equivalent rate.

Angle Measure (degrees)

distance (x)	θ_1	$\theta_1+\theta_2$	θ_2	$\theta_1+\theta_2+\theta_3$	θ_3	$\theta_1+\theta_2+\theta_3+\theta_4$	θ_4
0	90	90	0	90	0	90	0
0.25 D	75.964	82.875	6.911	85.236	2.361	86.424	1.18
0.50 D	63.435	75.964	12.529	80.538	4.574	82.875	2.337
0.75 D	53.13	69.444	16.316	75.964	6.52	79.38	3.416
1.00 D	45	63.435	18.435	71.565	8.13	75.964	4.399
1.25 D	38.66	57.995	19.335	67.38	9.385	72.646	5.266
1.414 D	35.268	54.74	19.472	64.764	10.027	70.531	5.767
1.50 D	33.69	53.13	19.44	63.435	10.305	69.444	6.009
1.75 D	29.745	48.814	19.069	59.744	10.93	66.371	6.627
2.00 D	26.565	45	18.435	56.31	11.31	63.435	7.125
2.25 D	23.962	41.634	17.627	53.13	11.496	60.642	7.512
2.449 D	22.212	39.237	17.025	50.774	11.537	58.523	7.749
2.50 D	21.801	38.66	16.858	50.194	11.534	57.995	7.801
2.75 D	19.983	36.027	16.044	47.49	11.463	55.491	8.001
3.00 D	18.435	33.69	15.255	45	11.31	53.13	8.13
3.25 D	17.103	31.608	14.505	42.709	11.101	50.906	8.197
3.464 D	16.103	30	13.897	40.894	10.894	49.107	8.2132
3.50 D	15.945	29.745	13.8	40.601	10.856	48.814	8.2127
3.75 D	14.931	28.072	13.141	38.66	10.588	46.848	8.188
4.00 D	14.036	26.565	12.529	36.87	10.303	45	8.13
4.25 D	13.241	25.201	11.96	35.218	10.017	43.264	8.046
4.50 D	12.529	23.962	11.433	33.69	9.728	41.634	7.944
4.75 D	11.889	22.834	10.945	32.276	9.442	40.101	7.825
5.00 D	11.31	21.801	10.491	30.964	9.163	38.66	7.696

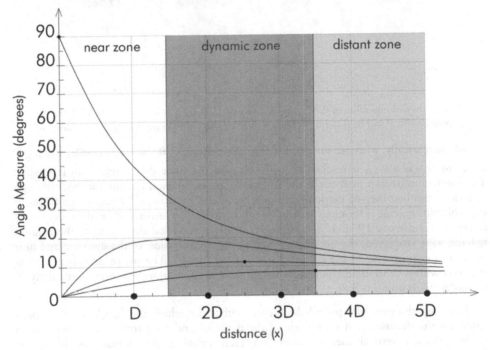

Fig. 6a (above) and 6b (below). Schedule of angle sizes for varying values of x; Θ_{max} as a function of the position of P. In the dynamic zone some view angles are increasing while others are decreasing

Returning to the Parthenon, let us apply the mathematics of the dynamic zone to the visual refinements, keeping in mind that the procession from Propylaea to Naos allows for both a lateral and frontal apprehension of the temple's form. If we now consider figure 5 and locate the columns of the Parthenon tangent to points adjusted to the distances between the columns, we can intuit that the reducing of any of the intervals between the columns will exaggerate the convex shield effect as we approach the temple causing the periphery to recede more quickly (fig. 7).

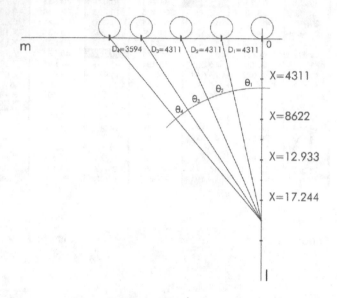

Fig. 7. Intercolumniation of the Parthenon: At $x = \sqrt{2} D$, θ_4 is at its maximum. With the shorter distance for D_4, P will be closer to O when θ_4 is at its maximum compressing the dynamic zone

Mathematically, when an interval D_n is smaller, the $\tan^{-1} \theta_n$ becomes smaller placing P closer to O when θ_{max} is reached as might be expected since the temple front is smaller. The intercolumniation intervals of the Parthenon are equal except at the corners, although variations can be observed particularly along the lateral elevations.[14] These irregularities in the column spacing Dinsmoor attributes to construction error rather than deliberate intention [Dinsmoor 1974: 178]. However, if the incremental decreasing of the column spacing were employed, the net effect would be the compression of the dynamic and near zones and a shortening of the duration of the visual "snap". The use of decreased spacing intervals within centralized colonnades would later become more common particularly in the orders with uniform friezes.

Inclining the peristyles provided another method by which the Greek builders could compress the duration and magnitude of the dynamic and near zones. Here they could retain constant vertical measurements, yet each visual angle measure would appear decreased in comparison to its complement on a perpendicular flat surface (fig. 8).

The slight backward tilting of the columns and entablature leads to a reduction of both the near and distant visual angles and causes the maximum view angles to be reached closer to the façade, thus shortening the duration of the shield effect (fig. 9).

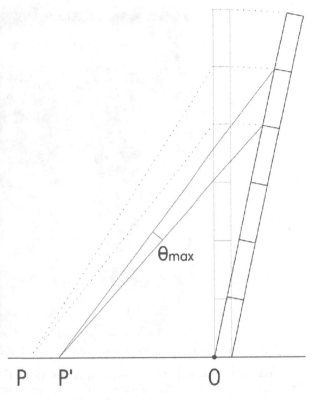

Fig. 8. The inclination of the
peristyle of the Parthenon. Photo by
the author

Fig. 9. With the inclination of the peristyle, θ_{max} is reached
closer to O. As a result, the dynamic zone is compressed

By drawing the far zone closer to the façade, the impression of the snapping back of the entablature is compressed. As with intercolumniation where the intervals D_n are incrementally decreasing from the center, or when the line m is inclined, as in the Parthenon, the mathematical expressions are more complex than (2) and (3) above. In both cases, however, the position of P at $\theta_{n(max)}$ moves closer to O reducing the domain and magnitude of the dynamic zone, which thereby reduces the acceleration of the snap.

Entasis accomplishes a variety of visual purposes and has, through the centuries, been accomplished in a number of ways from a simple tapering of the column shaft to an exaggerated bottle-shape found in Renaissance and Baroque buildings. The entasis of the typical Parthenon column is a simple linear taper from a base to a point about one-fifth the 10.5M height of the column at which point a gentle curve blends a slightly more inclined taper to the necking ring beneath the fillets (fig. 10).

Fig. 10. Column entasis of the Parthenon in the context of adjacent columns. Photo by the author

From any distant station point, especially from the most prominent view at the Propylaea, the entasis of the columns in the west and north peristyles is quite pronounced. The increased visual space between the columns at the top lightens the visual mass of the entablature while the complementary increase of the girth at the base firmly anchors them to the stylobate.

From a roaming point of view, entasis accomplishes two visual purposes, one involving perpendicular movement and the other parallel movement. Approaching the east peristyle perpendicularly, the upward taper of the columns facilitates the visual inclination of the façade, and helps to compress the dynamic zone. Ascending the stereobate and passing between the columns involves a movement where the viewer's position approaches and then recedes from the line of the columns. The impression is one of column sway. This phenomenon is especially pronounced along the extended path past the north peristyle where the processioner is moving parallel to the elevation. Here, the sway of the columns is directional, leaning forward as they are approached, swaying back into the vertical when alongside, and continuing their forward sway as they are passed. Again, their upward taper compresses the period and magnitude of the dynamic zone, buffering their visual movement, the "pushing" and "pulling" of the capitals.

Perhaps the most dramatic effect can be experienced in the downward curvature of the stylobate which is noticeably curved even when standing at a stationary position beside it (fig. 11).

Fig. 11. The curvature of the stylobate of Temple of Apollo at Selinute, Sicily. Photo by the author

The stylobate of the Parthenon rises 60mm at the center along the east and west fronts, and 110mm along the lateral faces so that from the center of the façade; the distant view angles are smaller as a result of the gentle downward curvature. In ascending the stereobate to enter the Pronaos, the upward "snap" of the corners is affectively countered, the result of a concave visual shield since in this case the viewer's eye is moving away from the surface plane of the stylobate. Simultaneously, the downward curvature of the entablature, most likely is the consequence of the curved stylobate supporting columns of equal length, counters the convex shield as the viewer ascends to the stylobate, slowing down the upward snap of the peripheral columns.

Conclusion

In our study we have assumed that the visual refinements were intended to counteract the geometry of visual space, acting as its inverse complement. We have noted some of the deficiencies of the linear and curved systems of perspectival representation to account accurately for the refinements, emphasizing that each of the methods rely on a stationary point to project a three-dimensional form onto a two-dimensional surface. As the formal system of the visual refinements was reinterpreted over the centuries, artists and commentators took often opposing positions as to the mathematical explanation of the tapers and curves that account for the optical corrections. While it is certain that the Greek builders intended to correct optical inconsistencies, and that the refinements provide evidence of a curved visual space is justified, the mathematics of that space remains speculative. As the art historian James Elkins writes, "Subjective curvature may be unique to the history of art in that it is a genuine unsolved scientific problem. It is the only case I know in which real, ongoing science enters into art history. The analytic question of subjective curvature is an unsolved scientific problem" [Elkins 1994: 186]. As with the many other accounts of visual space, Roaming Point Perspective offers a possible explanation for the geometry of visual space, and it too has its foundations in the Greek visual refinements.

Notes

1. [Vitruvius 1960: 89]. The curvature of the stylobate was accomplished by the use of leveling blocks of varying dimensions that incrementally decrease in "small steps" (*scamilli impares*) toward the center to generate the upward curvature of the surface; see [Panofsky 1991: fig. 11, 89].

2. [Alberti 1988: 189-218]. Throughout the first nine chapters of Book VII, Alberti prescribes many of the same refinements that were discussed by Vitruvius as necessary for visual harmony. Some 400 years later, Le Corbusier would cite the refinements of the Parthenon as the ultimate expression of a "plastic" architecture – a "pure creation of the mind" [Le Corbusier 1970: 185-207].

3. Helmholtz was especially critical of entire a priori justification of Euclidean geometry claiming it was inconsistent with the actual facts of perception; see [Körnigsberger 1906: 254-266].

4. For a concise explanation of Luneburg's mathematics of hyperbolic visual space, see [Heelan 1983: appendix, 281-319].

5. Scholars have not adopted universally accepted measurements for the Parthenon. Accordingly, I have relied on the measurements referenced in [Dinsmoor 1974].

6. Scholars have ascribed the origin of curvature of the stylobate with the functional necessity of accommodating the drainage of rainwater. This interpretation, while plausible, is not consistent with all temples, however. For example, to the east porch of the Erechtheum does not exhibit any curvature at all [Rhodes 1995: 76]. Another interpretation provided by Mavrikios [1974] claims the downward pitch at the corners of the stylobate and the entasis of the columns as a means of accommodating the visual weight of the temple. As he sees it, the Greeks employed principles of empathy to visually respond to the transference of the temple's gravitational weight to its foundation..

7. For a summary of the development of projective geometry from perspective, see [Klein 1970: ch. 14, 285-301].

8. [Perrault 1993: 162]. Perrault aims his criticism at the fact that distortion can only be corrected from a single point of view, leaving all other viewpoints compromised. "Even when the judgment of sight might not be able to prevent the distance and position of objects from deceiving us, the alteration of proportions is still not a good remedy for this supposed effect, because the effect of alteration is only at a given distance and only if the eye does not change position. There are optical figures whose proportions are modified in such a way that their effect is only favorable only if they are viewed from a specific location, these proportions, like those of optical figures, will also appear totally defective as soon as the viewer changes place, because an aspect that is oblique when a person is near becomes progressively less so as he moves away."

9.   It should be noted that the most exaggerated visual distortions in perspectival projection occur outside the normal field of vision and are therefore not readily detectable. Nevertheless, when following the methods of linear perspectival construction, these inconsistencies will appear within the field of vision as well as Leonardo shows.

10.   [Elkins 1994: 181-216]. Elkins provides a good commentary of the controversy surrounding curved perspective which includes an historical overview and examples of models that have been proposed.

11.   [Gombrich 1960: 258]. Gombrich states, "It is perhaps significant that the prime argument for this claim of a curvilinear world is taken from architecture and not painting. The Greeks allegedly introduced the so-called 'refinements' of deviation from rectangularity in their temples to correct the distortions of vision."

12.   [Heelan 1983: 29]. According to Heelan, "the relationship of the [Greek] refinements to the near zone of hyperbolic visual space is an intriguing one."

13.   [Panofsky 1991: 35-36]. Euclid's Proposition 8 states that "equal and parallel magnitudes at an unequal distance to the eye are not seen proportionally to the distances." Accordingly, Panofsky argued that Euclid recognized that scale is determined by view angle and not by distance thus countering the central premise of linear perspective and thereby providing support for curvilinear constructions.

14.   [Lawrence 1973: 173]. Lawrence includes a measured sketch of the north peristyle.

References

ALBERTI, Leon Battista 1988. *On the Art of Building in Ten Books.* Joseph Rykwert, Neil Leach and Robert Tavernor, trans. Cambridge, MA: MIT Press.

BROWNSON, C. D. 1980. Euclid's Optics and its Compatibility with Linear Perspective. Thesis? University of California at Santa Barbara.

DINSMOOR, William Bell. 1974. The Design and Building Techniques of the Parthenon – 1951. Pp. 171-199 in *The Parthenon*, Vincent J. Bruno, ed. New York: W W Norton.

ELKINS, James. 1994. *The Poetics of Perspective.* Ithaca, NY: Cornell University Press.

FLETCHER, Banister. 1967. *A History of Architecture on a Comparative Method.* 17th ed. New York: Charles Scribner's Sons.

GOMBRICH, E. H. 1960. *Art and Illusion: A Study in the Psychology of Pictorial Representation.* Princeton: Princeton University Press.

HEELAN, Patrick A. 1983. *Space Perception and the Philosophy of Science.* Berkeley: University of California Press.

KLINE, Morris 1970. *Mathematical Thought from Ancient to Modern Times.* Oxford: Oxford University Press.

———. 1954. *Mathematics in Western Culture.* Oxford: Oxford University Press.

KÖRNIGSBERGER, Leo. 1906. *Hermann von Helmholtz.* Frances A Welby, trans. Oxford: Clarendon Press.

LAWRENCE, A.W. 1973. *Greek Architecture.* 3rd ed. Middlesex: Penguin Books.

LE CORBUSIER. 1970. *Towards a New Architecture.* Frederick Etchells, trans. New York: Praeger Publishers.

LUNEBURG, Rudolph K. 1947. *Mathematical Analysis of Binocular Vision.* Princeton: Princeton University Press.

MAVRIKIOS, A. 1974. Aesthetic Analysis Concerning the Curvature of the Parthenon. Pp. 199-224 in *The Parthenon*, Vincent J. Bruno, ed. New York: W. W. Norton.

PANOFSKY, Erwin. 1991. *Perspective as Symbolic For.* New York: Zone Books.

PERRAULT, Claude. 1993. *Ordonnance for the Five Kinds of Columns after the Method of the Ancients.* Santa Monica, CA: The Getty Center for the Art and the Humanities.

RHODES, Robin Francis 1995. *Architecture and Meaning in the Athenian Acropolis.* Cambridge: Cambridge University Press.

SCULLY, Vincent. 1991. *Architecture: The Natural and the Manmade.* New York: Saint Martin's Press.

SERLIO, Sebastiano. 1982. *The Five Books of Architecture*. New York: Dover Publications.
VITRUVIUS. 1960. *The Ten Books of Architecture*. Morris Hickey Morgan, trans. New York: Dover Publications.

About the author

Michael C. Duddy is a founding principal of Designgroup International, an architectural practice with offices in New York and Shanghai specializing in urban mixed-use buildings. He is currently researching aspects of the history of the relationship between the built environment and the disciplines of epistemology and metaphysics. He is a guest critic at the School of Visual Arts in New York City.

Roberto B.F. Castiglia

Marco Giorgio Bevilacqua

Department of Civil Engineering, Architecture and Town Planning
University of Pisa
Via Diotisalvi 2
56126 Pisa ITALY
r.castiglia@ing.unipi.it
mg.bevilacqua@ing.unipit.it

Keywords: Turkish baths (Hammam); Elbasan; Albania; ornament; descriptive geometry; Euclidean geometry; proportional analysis; proportion; structures/structural engineering; dome; vaults; metrology

Research

The Turkish Baths in Elbasan: Architecture, Geometry and Wellbeing

Abstract. The two Turkish baths in Elbasan have been object of a remarkable survey as part of a vast international research. The original system of the two public baths dates back to the mid-sixteenth century. In the Albanian typology of the *hammam*, the various environments of increasing temperatures are located along a longitudinal axis ending in the cistern area where the boiler is located. The Albanian baths reveal their Ottoman roots in the complex concept of the vaulted structures. Different types of vaults in a same space are placed side by side in order to realize more complex compositions; the inner surfaces of the vaults are embellished by elaborate decorations that resemble stalactites. The complexity of the plan and the decorations reveal a rigorous geometric pattern that dominates the space and underlies its composition. All seems the product of a centuries-old wisdom where each element contributes to a complex system aimed at comfort and pleasure of the human being.

The architecture of hammam in the Islamic tradition

The tradition of the bath – *hammam* – was born and spread throughout Islamic society from the seventh century on, and performed a fundamental role up to the first half of the twentieth century. In addition to reasons linked to the care and cleansing of the body, the *hammam*, generally connected to mosques, represented the place where the user could complete the rites of purification imposed by the Islamic religion several times a day before prayer. The intimate pairing of the spheres of religion and comfort, and its availability to various levels of society, was determinant – in various regional forms – for the capillary diffusion of these structures in all Islam, so that some Moslems became true interpreters and successors of Roman and Byzantine thermal traditions.

A new interpretation of the Roman model is found in all the public baths in Islam. If they preserve in substance the systems of heating by hypocaust and by water adduction, the functional configuration of the environments appears to be substantially new. While the classical typology is organized along a spatial sequence formed by *apodyterium* (dressing room), *frigidarium* with swimming pool (*natatio*), *tepidarium*, *calidarium* and *laconicum* (steam bath), in the Islamic baths there is no *frigidarium* and swimming pool.[1] The *apodyterium* – the Turkish *cemekân*, the entry where the users could undress and relax after a thermal treatment – assumes a great symbolic importance.

The cemekân, which is generally square in shape and covered by a dome, is equipped on four sides by lofts, with benches and niches to guarantee guests' privacies. The entire environment is oriented around a fountain set almost exactly in the centre in order to preserve the centrality of the compositional genesis of the space.[2]

The *soğukluk,* the medium-temperature room, allowed the user to gradually get used to the rise in temperature during the passage from the unheated entry to the heated rooms. Directly connected with the latrines and the washrooms, it also functioned as a dressing room during the winter.

The *calidarium,* called *sicaklik,*[3] constituted the true heart of the entire spa. The smaller rooms for the steam bath – the *maghtas,* where temperature could exceed 40°C – are located over it. In several other models, the *gobektasi* – the "stone umbilicus" – is located at the centre of the *sicaklik;* this is a generally octagonal-shaped platform where the guests could lie down to relax or for a massage and shave.

The arrangement of the space mainly conforms to two general types of systems. In the first, typical of the *hammam* of the lower Western Mediterranean, the rooms are distributed along an axis that extends from the unheated entry through the rooms where the temperatures gradually increase and finishes with a cistern and boiler room.[4] In the second type, mostly developed on the Eastern Mediterranean, the distribution of spaces is less rigid, with polycentric models in which other rooms branch off from the central nucleus of the *sicaklik* which is surrounded by small exedrae.[5]

Although the types of *hammam* may differ according to the various regional traditions, they inevitably reflect purely technical rules of distribution, placing the higher-temperature rooms in direct contact with the technical rooms for the cistern and the boiler.

The Turkish baths at Elbasan

The city of Elbasan is situated in the Shkumbin river valley in central Albania. It was built on the remains of the ancient city of Skampa, a fortified centre founded in the first century B.C. on the Via Egnatia, the road which was the natural extension of the Appian Way and connected the harbour of Durazzo to the city of Byzantium. The historical city centre (*Kala*) makes the military origin of the location evident (fig. 1). In fact, the road system and the defensive perimeter of the ancient Roman fortress are preserved substantially in their original form today. An important commercial centre during the Ottoman domination, the city underwent significant interventions of modernization, establishing itself as a military outpost in the politics of expansion of the boundaries of the Ottoman empire towards the West.[6]

The two *hammam,* which represent two of the last existing models in the entire nation, are outstanding among the monuments built during the Ottoman period. The two structures were the subjects of an architectural survey performed by a general group of ten civil engineering/architecture students of the University of Pisa coordinated by Roberto Castiglia, teacher of Architecture Design of the Faculty of Engineering.[7] The surveys of the two baths, together with other monuments of the city of Elbasan, are part of the MIUR 2006-2007 research project entitled "Pilot scheme for knowledge, conservation and improvement of Kala Fortress in Elbasan, Albania" coordinated by Prof. Roberto Pierini of the University of Pisa.

Fig. 1. The *Kala* fortress of Elbasan, Albania. In evidence, the most important monuments: 1) "Shen Maria" Orthodox Church; 2) "Xhamia Mbretërore" Imperial mosque; 3) Kala's *hammam*; 4) "Sahati" clock tower; 5) Baazar's *hammam*

The first bath, within the fortified perimeter of the *Kala*, is set on the eastern side of the axis of the road that today marks the ancient *decumanus* of the Roman fortress. The second is found outside the walls, in the place where the ancient Bazaar was located and next to the mosque of "Agajt", to which the bath was closely connected. The first reports regarding the presence of the two structures date back to the second half of the twelfth century,[8] although it is possible to attribute their construction to the century before that.

The two *hammam* show plan systems quite similar one another, with the various rooms located along an axis. The bath of the Bazaar preserves in large part the characteristics of the original system, and underwent several restoration interventions between 1973 and 1977 [Shtylla 1979: 83-90]. The main spaces are identified in fig. 2: The great entrance room – the *cemekân* (1) with the central well and the wooden gallery – leads to the vaulted medium-temperature room (2), the *soğukluk* (3) with the washing room, and the latrines (4, 5). Access to the *sicaklik* (6, 6a) was originally possible exclusively from the intermediary room. The direct connection between the entry and the room was in fact opened during the restoration works of last century. Finally, from the heated room are accessed the two *maghtas* for the steam baths (7a, 7b), placed side-by-side and adjacent to the cistern (8) with the boiler (9).

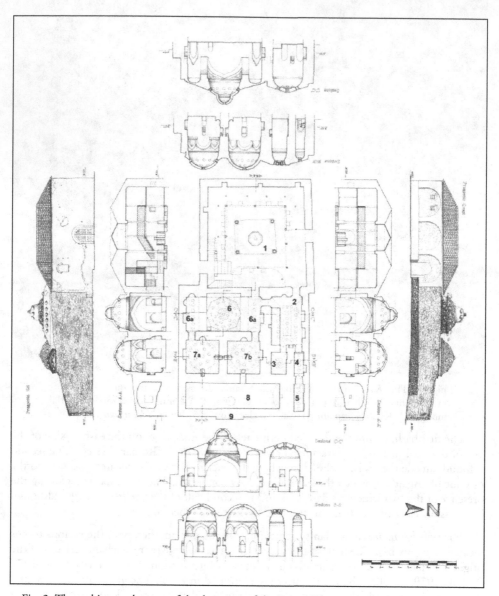

Fig. 2. The architectural survey of the *hammam* of the Bazaar. Plan, vertical sections and facades.
Drawing by A. Cartei, F. Degl'Innocenti, J. Farsetti, A. Ferrara

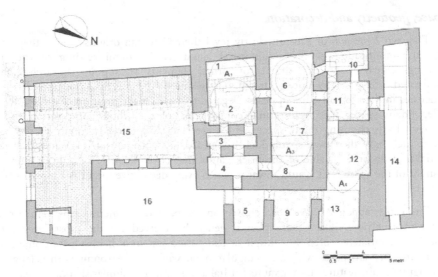

Fig. 3. The architectural survey of the *hammam* of the *Kala*. Plan. Drawing by M.G. Bevilacqua

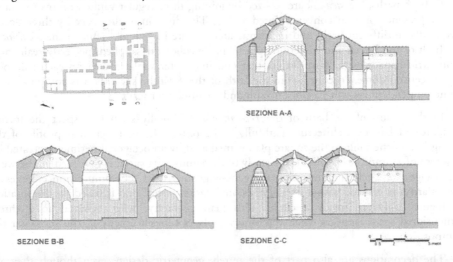

SEZIONE A-A

SEZIONE B-B   SEZIONE C-C

Fig. 4. The architectural survey of the *hammam* of the *Kala*. Vertical sections. Drawing by M.G. Bevilacqua

The plan of the baths of the *Kala* is more complex (fig. 3). It is possible to identify at least two constructive phases. The typological characteristics of the bath with the axial distribution of the spaces can be seen in the central nucleus comprising the *soğukluk* at the entry (1, 2) with the attached sanitary fittings (3, 4), the *sicaklik* (6, 7, 8), the steam bath area (10, 11, 12) and finally the cistern with the water heating systems (14). The second phase of construction to extend the baths can be probably be identified in the total rearrangement of the *cemekân* (15, 16) and in the construction of the two heated rooms (6, 9) reserved for women.[9]

Space, geometry and decoration

The two *hammam* of Elbasan clearly revel their Ottoman origins even if they show a general simplification of the architectural language. A careful reading of the survey drawings clearly shows that the set of formal elements making up the architecture is dominated by a purely geometric design. The geometric genesis does not pervade the two structures in their entirety, but appears relegated to the formal definition of each individual, independent space. The architectural composition of the whole composition therefore proceeds by juxtaposing many spaces which are formally enclosed. Square or rectangular spaces each have their own spatial identity, created almost exclusively by attention to the spatial and decorative value of the brick vaults that cover them. The only characterizing element of the space is the vault, towards which every decorative intent is addressed (figs. 2 and 4).

The vaults appear to have been chosen on the basis of geometric-constructive reasons and representations. The square plan is generally reserved for the most representative spaces, appropriately roofed by a simple or compound system of hemispheric domes. Barrel vaults are used in spaces with a rectangular plan, while in the rooms with a lengthened rectangular shape feature cloister vaults (in Italian, *a schifo*) or similar shapes.

In both baths, the *sicaklik* are realized by joining three regular vaulted rooms mediated by the presence of great convex pointed arches. The division of the space by three occurs frequently in different baths and in Islamic architecture in general.[10] But in the *sicaklik* of the bath of the Bazaar the trisection appears more coherently applied, perfectly realizing a symmetrical space, with barrel vaults that end in cloister vaults placed on each side of a higher central dome. In contrast, in the bath of the *Kala* the succession is composed of a dome on pendentives, with a cloister vault and a half-cloister vault.

In the rooms of the bath of the *Bazaar*, which actually seem to respect the formal standards of Islamic architecture faithfully,[11] the passage from the circular profile of the spring line of the vault to the square plan is mediated by an octagonal perimeter created by a system of *trompe*. It is really in the study of the connecting surfaces that mediate between the octagon or circle squaring, that the genesis geometric of the composition emerges in total clarity. It is made with a two-dimensional design system and is based on the rotation of regular polygons inside a circle. This plan geometry, when referred to a three-dimensional space, appears when the observer places himself exactly at the centre of the composition.

The decorations are also part of the purely geometric design, even though they are limited to the intrados of the vaults. Especially in the domes, the decoration conforms to the formal technique used to set the vaults on the walls, thus increasing the geometric complexity of the surface that joins the circle to the square, in a composition which is comprehensible only when it is projected onto the plan.

In the *maghtas* of the bath of the Bazaar and in the small internal room of the bath of the *Kala*, honeycomb-like shapes are found in the decorations of the vaults. These are similar to the most complex decorative technique of *muqarnas* but do not achieve the complexity and the stereometrical wealth typical of the most mature examples of Islamic architecture.[12]

A series of oculi, *jamat*, embellishes the intrados surfaces of the vaults and is in keeping with the vault geometry: along the parallels in the hemispherical surfaces, along the generating lines in the cylindrical surfaces or in the top of the cloister vaults. Generally round or hexagonal in shape, the oculi provide the only source of light inside the rooms.

However, regards for formal decoration appears on the interiors. Outside, the structures appear bare, the building masses deprived of all architectural pretension. The only meaningful element is given by the silhouette of the vault, which declares to the outside world the compositional genesis "by juxtaposition of the spaces" and is the only element that marks the two monuments in the urban scene.

The geometry of the arches and vaults

The two *hammam* of Elbasan are characterized by arches and vaults that fall well within the tradition of the Ottoman architecture. While not representing paradigmatic examples in the international culture, they constitute a significant patrimony for Albania. Further, it should be underlined that Elbasan's Turkish baths of the Ottoman period have not been object of in-depth studies. As regards the different typologies of the arches and vaults used in the structures under examination, the geometric analysis proposed here represents only a synthesis of the knowledge acquired during the architectural survey. For this reason, this present study refers to the arches and vaults located in the *hammam* inside the *Kala* (figs. 3 and 4). This synthesis does not consider analogous structures that have been surveyed in other monuments of the fortified citadel of Elbasan, among which the orthodox "Shen Maria" church and, with reference to the arches, the mosque "Xhama Breterore" are worth to be mentioned.

Measurements carried out on arches A2 and A3 indicate an ogival pointed or inflected arch whose development is found within a segmental arch (fig. 5). For the two arches under examination a very slight angular point can be observed; as a result, the centers of the two arches that converge at that point do not line up.

The geometric characteristics of arches A1 and A4 refer to the Persian arch surveyed by Dieulafoy-Choisy (cf. [Docci and Migliari 1992]). Here the distance from the center of the arch that precedes the angular point to the center of the clear span is equal to 1/8 of the length of the span itself. The overlay of the geometric construction of the Syrian arch with the arches surveyed has shown a significant difference in the case of arches A1 and a substantial correspondence in case of arch A4, except for a slight alteration in the right profile. In case of arch A1, after having reduced by successive attempts the construction of the Syrian arc for a quota equal to 1/12 of the span, a good correspondence was obtained. It should be noted that the imposed drop corresponds to the fraction which in the construction of the Syrian arch brings to the graphic determination of the two centers corresponding to the arches that from the angular points arrive to the vertex of the ogival pointed profile.

As regards arches A2 and A3, the measures carried out show an ogival pointed profile. Arch A2, as well as constructive imperfections, has been identified in a pointed arch at the fourth point, but with the centers reduced by 1/10 of the clear-span. The profile of arch A3 does not diverge from a pointed curvature at the third point with centers dropped by 1/10 of the span, as is the case for arch A2.

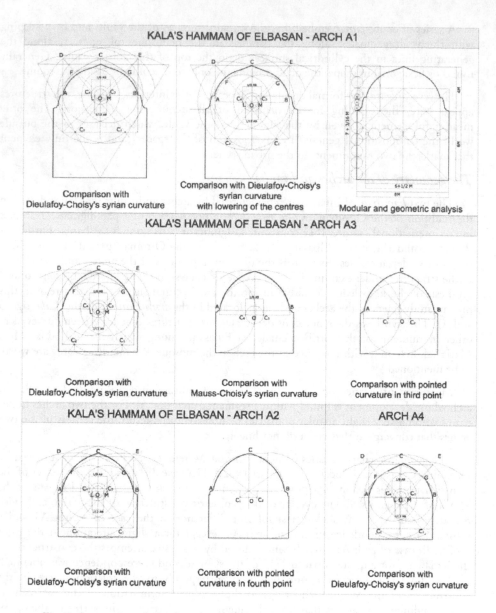

Fig. 5. Geometric analysis of curvatures in the arches of the hamman of the *Kala*, Elbasan. Drawing by R.B.F. Castiglia

An over-arching geometric analysis of the system arch-cornice-abutment with reference to arch A1 brought to light the following conclusions:

a) the relationship between the height of the vertex of the arch and the clear span corresponds to 10/(6-1/2);
b) the relationship between the height of the vertex of the arch and the opening light is 10/8;
c) the height of the cornice is 1-3/16.

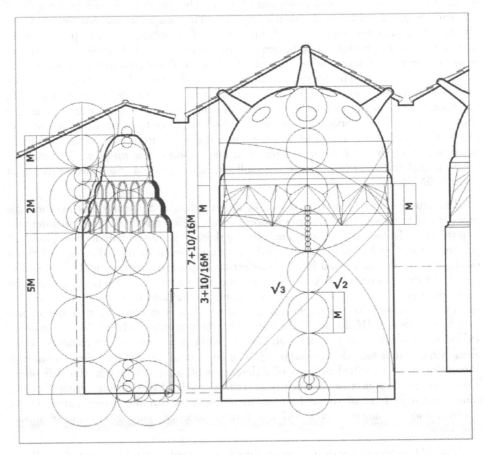

Fig. 6. Geometric and modular analysis of the vaults 10 and 11 in the *hammam* of the *Kala*, Elbasan. Drawing by R.B.F. Castiglia

The vaults covering the rooms of the two *hammam* are described in their essential characters in tables 1 and 2. The barrel vault with number 2, corresponding to the *soğukluk* of the *hammam* of the *Kala* is set on circular profile. The link between the walls and the impost provided by spherical pendentives deformed in such way so as to connect the perimeter of the impost of the dome to the circumference. The intrados of the spherical pendentives are decorated with *muqarnas* that play no structural role (figs. 3 and 4). The dome, which according to the measurements corresponds to a semi-ellipsoid, is pierced by

13 oculi, to be interpreted as intersections of the dome with right circular cones whose axes do not converge in an unique and significant point.

Vault number 10, which is found in one of the three *magthas* of the *hammam*, corresponds to the type *a schifo*, that is, a barrel vault ending in cloister vaults with an elliptic curvature and cut by a surface that is basically flat. There are four oculi inserted in the plain part of the vault; these are conic surfaces with axes perpendicular to the room flooring and parallel one another. The vault is set on two frames, with flaring toward the inside that determine a general projection of 4 cm. The frames are joined to the walls of alveolar *muqarnas* with simple sketch (fig. 8). The *muqarnas* realized in terracotta alternatively foresee the composition of three orders of niches staggered and projecting the one on the other determine a middle general projection of 35 cm. The tallest and external points of every line of niches belong to an arch of circle with the convexity turns toward the outside of the room. The transversal dimension of the room is 145 cm, the smaller diameter of the ellipse section of the vault, in such direction, is reduced to only 75 cm. The relationship between the width of the room and the diameter of the ellipse is therefore about 2 (fig. 6). The relationship between the ideal height of the vault (without cut with the horizontal plan) and the width of the room results equal to 3. Assumed as module M equal to half width of the room, it is observed that impost of the *muqarnas* is found to a height of 5M and that the impost of the vault corresponds to 7M and therefore its higher point to 9M.[13]

The vault number 11 is characterized by a hemispherical dome and presents the same number and typology of oculi in comparison to vault 2 (figs. 3 and 4). The link among the impost circumference of the dome, mediated by three ring frames attaining a height of 27 cm and project of 8 cm, is characterized by a composition in three-dimensional forms similar to crystals, determined by the intersection of prismatic surfaces and conic and/or fluted surfaces for a height of 70 cm (K) and set to the quota of 275 cm (L) in comparison to the floor plan. Such quota of impost corresponds to the diameter of the dome (279 cm). Assumed as module M the height corresponding to K it is observed that the impost corresponds to 3+10/16 M, while the top of the vaults reaches 7+10/16 M. The median section of the dome covered room corresponds, without the frames, to a rectangle whose relationship among the sides is equal to $\sqrt{3}$. The interspaces between the four series of such compositions are identified by bows of different profile around four sides and left plane. The orthogonal projection of such geometries reveals that the four groups of the compositions in form of crystal lean on an octagon, so that the determination of the circumference of dome impost includes further spherical joining of modest surface extension.

It should be noted that the octagon is found in the vaults of the *hamman* of the *Bazaar* (fig. 2) with a different constructive characterization. The regularity that marks the functional spaces of the *hammam* of the *Bazaar* is not found in the Turkish bath located inside the citadel. This condition is transferred to the different geometry of the vaults. In the three domes where an octagonal layout is observed, this is concretized by more frames sustained by *trompe* or ogival niches to the sides of the lateral arch, also with an ogival profile.

The orthogonal projection of the said compositions as for the *muqarnas* of vaults 2 and 10 of the *hamman* of the *Kala*, shows an aspect of a certain interest and faced, for instance, from I. I. Noktin (1995) about the possibility to move the two-dimensional representation to three-dimensional constructions of a certain complexity. In the case of the *hammam* of Elbasan it seems reasonable to affirm that, more than a rigorous transposition of the sketch, the construction has seen an empirical application of elements founded on operational criterions codified by local workers.

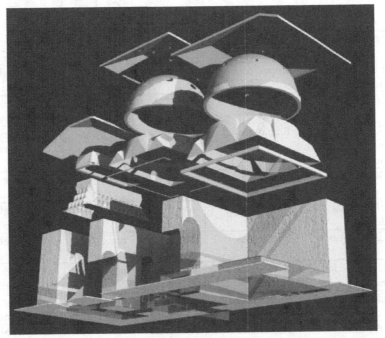

Fig. 7. The architectural survey of the *hammam* of the *Kala*, Elbasan. 3D digital model by V. Bertola, V. Cardini, A. Cartei, A. Miano, G. Romeo, G. Simonelli, D. Tofanelli and C. Verdini

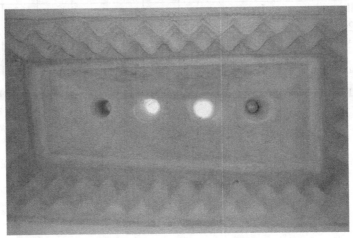

Fig. 8. *Muqarnas* decorations in the *hammam* of the *Kala*, Elbasan

n	TYPE OF VAULT	PLAN OF THE ROOM	PENDETIVES	TROMBE	MUQARNAS	FRAMES	OCULI	
							NUM.	TYPE
1	"A schifo"	Imperfect rectangles	no	no	no	yes	2	Square pyramid
2	Ellipsoidal dome	Imperfect square	yes	no	yes	yes	13	Right circular cone
3	"A schifo"	Rectangles	no	no	no	yes	4	Hexagonal pyramid
4	Barrel vault with cloister heads	Rectangles	no	no	no	no	4	Hexagonal pyramid
5	Barrel vault	Imperfect rectangles	no	no	no	no	8	Right circular cone
6	Ellipsoidal dome	Imperfect square	yes	no	no	yes	13	Right circular cone
7	Cloister vault	Rectangles	no	no	no	yes	12	Right circular cone
8	Half cloister vault	Rectangles	no	no	no	yes	3	Hexagonal pyramid
9	Ellipsoidal dome	Imperfect square	yes	no	no	yes	9	Right circular cone
10	"A schifo"	Rectangles	no	no	yes	yes	4	Right circular cone
11	Ellipsoidal dome	Imperfect square	no	yes	no	yes	13	Right circular cone
12	Ellipsoidal dome	Imperfect square	no	yes	no	yes	13	Right circular cone
13	Barrel vault	Rectangles	no	no	no	yes	8	Right circular cone
14	Barrel vault	Rectangles	no	no	no	no	-	-

Table 1. Synthetic analysis of the vaults in the *Kala*'s hammam of Elbasan, Albania

n	TYPE OF VAULT	PLAN OF THE ROOM	PENDETIVES	TROMBE	MUQARNAS	FRAMES	OCULI	
							NUM.	TYPE
1	-	-	-	-	-	-	-	-
2	"A schifo"	Rectangles	no	no	no	yes	20	Hexagonal pyramid
							3	Star-form pyramid
3	Cloister vault	Rectangles	no	no	no	yes	6	Hexagonal pyramid
							1	Star-form pyramid
4	"A schifo"	Rectangles	no	no	no	yes	5	Hexagonal pyramid
5	"A schifo"	Rectangles	no	no	no	yes	5	Hexagonal pyramid
6	Hemispherical dome with lantern	Square	no	yes	no	yes	24	Hexagonal pyramid
							1	Star-form pyramid
6a	Schifo	Rectangles	no	no	no	no	6	Hexagonal pyramid
7a/b	Hemispherical dome	Square	no	yes	no	yes	18	Hexagonal pyramid
							1	Star-form pyramid
8	Barrel vault	Rectangles	no	no	no	no	-	-

Table 2. Synthetic analysis of the vaults in the *Bazaar*'s hammam of Elbasan, Albania

Conclusions

These observations are based on the results of the architectural survey carried out in 2006/2007 in the city of Elbasan, especially in the fortified Ottoman citadel of the fifteenth century (the *Kala*), once a Roman Castrum. The three-dimensional modeling of the most significant monuments of the *Kala*, partly discussed in the present article and as yet not published in its entirety, constitutes a contribution to the knowledge of those structures in light of their maintenance and use. From this point of view the present publication is aimed at promoting further initiatives in other cities of Albania in order to develop the knowledge of the monuments.

Notes

1. The reason there is no swimming pool in the public baths is due to the Koran, which prescribes a complete washing of the body by immersion exclusively after contaminations and the sexual relations. Before prayer, the believer must generally wash his face, then his arms up to his elbows, then rub his head and his feet up to his ankles. Nevertheless, a basin for immersions is found in some healing baths as those in Cefalà Diana in Palermo, Sicily, or Khirbat al Majar near Jericho [De Miranda 2005: 344]..

2. The *cemekân*, also known as *maslah* [Micara 1985: 114-15] was often used as a reception room where the authorities organized parties with music and dancing.

3. The *calidarium* is also known as *bayt al-harara,* while with *bayt al-awwal* can be identified as the the *tepidarium,* which is referred to in the text as *so ukluk* [Micara 1985: 114-15].

4. Recall the Moroccan *hammam* "al-Mokhfiya" in Fès from the first half of the fourteenth century or those of "Chelia" and "el-Alou" in Rabat, and the Alhambra bath in Spain, all in the first half of the fourteenth century [De Miranda 2005: 350-353].

5. For instance refer to the baths "Sitti Adbra", "al-Buzuryya" and "al-Zan" in Damasco respectively of the twelfth, thirteenth and fourteenth centuries [De Miranda 2005: 350-353].

6. In fact, in 1466 the fortification of Elbasan was reconstructed by the Ottoman sultan Mehmet II on the ancient Roman plant Roman in the configuration that is seen today.

7. Architectural Survey of the *Hammam* of the Kala, April 2006: Professor: R.B.F. Castiglia; collaborators: M.G. Bevilacqua, B. Ruggieri, L. Salotti; students: F.Cinquini, D. Gemignani, N. Pagni, S. Turchi; Architectural Survey of the Hammam of the Bazaar, March 2007: Professor: R.B.F. Castiglia; collaborator: M.G. Bevilacqua; students: A. Cartei, F. Degl'Innocenti, J. Farsetti, A. Ferrara. The 3D graphic digital processing of the Turkish bath of the Kala was undertaken with the assistance of V. Bertola, V. Cardini, A. Cartei, A. Miano, G. Romeo, G. Simonelli, D. Tofanelli and C. Verdiani, students of the Laboratory for CAD Applications, first degree technical courses in Civil Engineering-Architecture at the University of Pisa; professor: M.G. Bevilacqua.

8. The presence of the two structures is given by Turkish Evlia Çelebiu in the chronicles of his 1672 voyage [Shtylla 1979a: 83].

9. Both sexes were permitted to use the bath of the *Kala* at the same time, unlike the *hammam* of the Bazaar, which was reserved for women on Thursdays at the last.

10. The successive and symmetrical partitioning for three architectural spaces is frequently found in other Turkish baths in particular and in Islamic architecture in general. According to Creswell, it would have originated in the housing units of the *omayyadi* buildings of the eighth century [Creswell 1966: 185-192].

11. A deeper analysis of the results of the survey shows some differences between the two baths. In particular, the bath of the *Kala* does not show the same geometric and constructive precision observed in the other *hammam*. A study still in progress may justify these "deformities" on the basis of the hypothesis, not still verified, that the *hammam* of the *Kala* was built on the remains of or as an amplification of more ancient thermal structures of Byzantine or perhaps Roman origin.

12. The term *muqarnas* refers to the stalactite-like decorations of the domes, very characteristic of Islamic architecture. The decorative system consists in arch-shaped overlapping projections of prismatic cells. They can be very hollow and formed the keystones leaning vaults, hence the name "stalactites".

13. It should be noted that the geometric relationships that refer to the floor level can be voided by the possible prominence of the same plane.

Bibliography

BIONDI, Benedetta, ed. 2005. *Architectural heritage and sustainable development of small and medium cities in South Mediterranean regions. Results and strategies of research and cooperation.* Pisa: ETS.

BRUE, A. 2003. *Cathedrals of the flesh. My search of the perfect bath.* London: Bloomsbury.

ÇELEBIU, E. 1930. *Shqipëria para dy shekujsh.* Tirane.

CERASI, M. 1986. *La città del Levante.* Milano: Jaca Book.

CLÉVENOT, D. 2000. *Decorazione e architettura dell'Islam.* Firenze: Le Lettere.

CRESWELL, Keppel Archibald Cameron. 1966. *Architettura Islamica delle origini.* Milano: Il Saggiatore.

DE MIRANDA, Adriana. 2005. From the thermal baths to the hammam. Pp. 343-345 in *Architectural heritage and sustainable development of small and medium cities in South Mediterranean regions. Results and strategies of research and cooperation,* Benedetta Biondi, ed. Pisa: ETS.

DOCCI M.- MIGLIARI R.. 1992. *Scienza della rappresentazione.* Roma: La Nuova Italia Scentifica.

GRABAR, O. 1989. *Arte Islamica. La formazione di una civiltà.* Milan: Electa.

HATTSTEIN, M., P. Deluis. 2000. *Islam Arte e Architettura.* Köln: Konemann.

HILLENBRAND, R. 1994. *Islamic Architecture.* Edinburgh: Edinburgh University Press.

HOAG, J.D. 1978. *Architettura Islamica.* Milan: Electa.

INZERILLO, M. 1980. *Le Moschee di Mogadiscio.* Palermo: Renzo Mazzone Editore.

MICARA, L. 1985. *Architetture e spazi dell'Islam. Le istituzioni collettive e la vita urbana.* Rome: Carucci Ed.

NOTKIN, I.I. 1995. "Decoding Sixteenth-Century Muqarnas Drawings". *Muqarnas* Volume XII: An Annual on Islamic Art and Architecture. Leiden: E.J. Brill.

SHTYLLA, V. 1979a. Restaurimi i dy banjave mesjetare në vendin tonë. *Monumentet* 17: 83-95.

———. 1979b. Banjat e mesjetës së vonë në shqipëri. *Monumentet* 17: 119-137.

STIERLIN, H. 1997. *Islam. Da Baghdad a Cordova. Architettura delle origini dal VII al XIII secolo.* Köln: Taschen.

———. 1999. *Turchia. Dai Selgiuchidi agli Ottomani.* Köln: Taschen.

VOGT-GÖKNIL, U. 1965. *Architettura ottomana.* Milan: Il Parnaso Editore.

ZANGHERI, L. ed. 1986. *Architettura islamica e orientale.* Florence: Alinea Ed.

About the authors

Roberto B.F. Castiglia earned his degree with honors in 1991 in Civil Engineering at the University of Pisa and his Ph.D. degree in 1999 at the Polytechnic of Turin. He's a researcher at the Department of Civil Engineering of Pisa, where his research activity is principally in the field of the Architectural and Urban Survey. He's a professor of Architectural Drawing in the degree program in Civil Engineering-Architecture at the Faculty of Engineering of Pisa University. Among his works: *I disegni degli ingegneri della Camera di Soprintendenza Comunitativa di Pisa* (Pisa-Roma 2001); *Chiesa e Badia di San Savino a Montone. Il rilievo* (Pisa 2001); *Centralità e uso del suolo urbano. Lucca* (Lucca 2007).

Marco Giorgio Bevilacqua earned his degree with honors in 2003 in Civil Engineering at the University of Pisa. Since 2004 he has been working on his doctorate in "Science and Techniques for Civil Engineering at the same university. His thesis is entitled "La fortificazione della città di Pisa nel XVI secolo. Il primo fronte bastionato". He is part of various national and international research projects on the subjects of the designing and surveying of architecture. He collaborates in teaching projects for Architectural Drawing in the first years of the degree program in Civil Engineering-Architecture at the University of Pisa. He's a professor of the Laboratory for CAD Applications in the same degree program. He works also as engineer.

Anat David-Artman

Kleinman 11, flat 12
Jerusalem 96552 ISRAEL
anatd28@gmail.com

Keywords: Vitruvius, harmony,
proportional systems, relational
systems, modular systems,
vitalism, Hans Driesch,
individuation, equipotentiality

Research

Mathematics as the Vital Force of Architecture

Abstract. This article shows that mathematics serves as a vital force in architecture. By comparing the characteristics that Hans Dreisch attributes to the vital force with the characteristics of mathematical proportions in architecture, this paper demonstrates that both can be seen as principles of individuation, both can be the source of equipotentiality, and both are manifested in harmony.

Introduction

Mathematics is usually considered a device of abstraction from every living, and therefore changing, aspect of reality. Considering architecture from the point of view of mathematics is usually understood as turning it into an abstract entity, eternal and frozen. Contrary to that conception, I claim in this paper that in architecture mathematics is used as a vital force. In spite of the common conception of mathematics as contrary to the living aspect of things, the association of mathematics with life is not totally absent, either in general or in architecture. For example Sir D'arcy Wentworth Thompson (1860-1948) uses mathematics not as something that freezes life but as a tool for analyzing its dynamic aspect. In architecture this association exists in two versions: one is the mathematical interpretation of "organic unity", for example in Alberti's theory, and the other is the mathematical analogy between man and column found in Vitruvius and his disciples. However, here this association will be presented in a somewhat different way. I connect it to a theory of life called "vitalism", and especially to that theory as presented in the writing of the biologist Hans Driesch (1867-1941).

According to Driesch, life cannot be fully explained by material rules. The rules of matter explain only the material aspect of the living organism, not its form. In order to explain the latter aspect, Driesch and the Vitalists turn to a principle which is not reducible to matter. This principle Driesch calls the "vital force". He says that this is a principle of individuation, that the evidence for its existence is the harmony that is found in organisms and that they are characterized by equipotentiality [Driesch 1908, 1: 76-149; 2: 310-318]. What I would like to demonstrate here is that in the case of architecture, mathematics is meant to be a principle of individuation of matter and that it is expressed in harmony and in equipotentiality.

The "vital force" as a principle of individuation

Driesch sees the vital force as a universal principle which by its embodiment in matter turns a piece of matter into an individual. The individual identity of this piece of matter from which the organism is made, which distinguishes it from the rest of the material in the world, does not depend on a material identity but on its vital force. In order to clarify this point, one may compare the organism to an inanimate object – for example, this cup I hold in my hand. It is not identical to another cup which stands on my desk although they look the same. That is, if I were to switch between them and hold the other cup, we would say it is not the same cup. It is not identical because it is not made from the same piece of matter. Therefore the identity of the cup depends on a material identity. When we say that this is the same object, we mean that its material identity is enduring. Let us say that in the first cup there is some coffee. We will say that it is the same coffee only if its material identity endures. If, on the other hand, I drink the coffee and pours more coffee into the same cup, we would say it is not the same coffee that was there before.

In contrast, in the organism the matter changes all the time but nevertheless we say it is the same organism. For instance, the identity of a cat does not depend on persistence of the matter it is made of, but on the contrary, on the constant replacement of this matter. The cat is exactly the same cat as long as its matter changes, but when it stops changing, the cat dies and in a way it is no longer a cat. Its individuality dissolves. The piece of matter that is what remains of the cat decomposes and once again becomes one with all earthly matter.

According to the Vitalists, what makes the cat a cat is some sort of vital principle (force) which uses matter but is external to it. This means that the cat stays the same cat in spite of its lack of material identity because of a principle, and not because of something material.

Equipotentiality as a testimony to the existence of the vital force

According to Driesch, one testimony to the existence of a vital force in the organism is what he calls "equipotentiality". Equipotentiality means that any piece of matter in the organism (at least in the early phase of its development) contains the potential to obtain the form of the whole organism. Driesch arrived at this conclusion after carrying out an experiment on a sea urchin, in which he eliminated half of the egg's cell. At the beginning, the cell evolved in the form of half its usual form but later it changed into an organism with a complete form. The same tendency was evident in a low sort of organism, clavellina. Driesch had cut only a part of it. At first the part lost its form, but then it organized itself into a new, smaller organism. He did another experiment, with a sea rose named tubularia, which has the form of a stalk with a head. Driesch observed that wherever he cut the stalk it grew a new head. What struck Driesch was that the proportion between the stalk and the head was always similar (remained stable). From all these findings Driesch concluded that the organism is an equipotential system, that is, every part of it has an equal potential to evolve into any part, or into the whole.

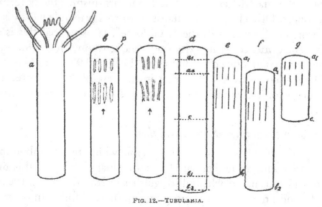

FIG. 12.—TUBULARIA.

Tubularia, from *The Science and Philosophy of the Organism* [Driesch 1908]

Harmony as a testimony to the existence of the vital force

Driesch attributes to the organism three kinds of harmonies. By the concept of "harmony", he means an inexplicable correspondence between different chains of events, which are not related causally. The famous example of harmony is the correspondence between different watches. In any watch there is a system of cog-wheels which operate on one another and create a chain of events that is responsible for the fact that the watch can show the time, but there is nothing that connects the different watches internally. What makes them show the same hour is a preestablished harmony put into them by the watchmaker who set them for the same time, and who is, of course, external to them. In the same way the Vitalists believe that the harmony between the different events that bring about the existence of the organism is put into them by an external vital force.

Driesch calls one of the three harmonies the "harmony of constellation". By this he means the way different organs are arranged in correspondence with each other and in a proportional relation, so that in the end a whole organism is created despite the relative independence of all the events that bring it about from the other events.

Mathematics as a principle of individuation of architecture

Modular and relational systems. According to Vitruvius, "Without symmetry and proportion there can be no principles in the design of any temple; that is, if there is no precise relation between its members, as in the case of those of a well shaped man" [1960: 72]. This famous mathematical analogy between architecture and man can be understood in two ways. One interpretation is that the proportion of a building should be similar to the proportion of man (for example, the number of times the head goes into the body is the same as the number of times the column's capital goes into the shaft); the second interpretation is that the similarity between them is in the very fact that in both there is a proportional relation between part and whole.[1] The different interpretations lead to two manners in which mathematics can serve as a principle of individuation of a building. On a lower level, the building receives an individual quality by its similarity to man, who is himself an individual. At this level the building does not become an individual but only resembles one. In a higher level, the similarity between the building's proportions and those of man is expressed by their functioning as a vital force.

As already stated, Driesch claimed that the vital force is universal but serves as a principle of individuation of matter. The individual is created by the union of the vital force with matter. In the same way we can say that in architecture mathematics is the principle of individuation of matter. Mathematics is, of course, something universal but by imposing a system of proportion upon a piece of matter, that piece becomes an individual that is different from its surroundings. This happens because proportions frame the object and force upon it some sort of ideal being.

Proportions are a sort of framing device because through them every part is related to the whole, and thus the system closes the object upon itself. They therefore relate the body only to itself. That is, it makes it possible to measure something without recourse to any external measure (a meter, for example). In other words, when we consider the proportion of the object no importance is attributed to the actual size of it, but only to the internal relation between the parts and their relation to the whole. This gives the object a sort of autonomy from its surrounding: it becomes an individual.

Using a system of proportion instead of actual sizes enables Vitruvius to determine canons of building typess (for example, temples) without recourse to a specific program or a specific site. Thus he establishes an ideal which is universal. However, imposing this ideal on a specific matter (connected to a specific program and site) turns this piece of matter into an individual. In this way the matter itself receives a certain "idealness" which differentiates it from its surroundings; on the other hand, by being embodied in a piece of matter, the ideal becomes an individual. Only the union between the ideal and the actual matter creates an individual building, in the same way that only the embodiment of the vital force in matter creates an individual organism.

We can now look at how individuation happens in different proportional systems. In general, we can divide the proportional systems into two main kinds: we will call them "modular" and "relational".[2] The first refers to those systems that are based on a module, i.e., every part of the body is measured by a standard module which might have its origin in one part of the body, but which may not necessarily be related to the other parts. Therefore this module might be projected on the body no less arbitrarily than a standard meter. The difference is that the module has no definite size that relates it to the actual world. Visually, modular systems can be presented as a homogenous grid into which the object is inserted in a certain habitual way. For example, we can find the use of this method in the technique of painting through a framed net that was presented in an etching by Dürer, or in the technique for producing sculptures in ancient Egypt, where they were constructed by means of vertical projections that were drawn into grids on the different sides of the stone. The art historian Erwin Panofsky (1892-1968) pointed out the fact that the parts of the body do not necessarily begin or end with the lines of the grid, but the grid is placed on them arbitrarily and thus mechanically [1983: 86]. Actually modular systems are not uniquely related to the body but are something one can put onto the body and the world continuously. The grid can continue endlessly and be projected onto anything outside the body as well as onto the body itself.

The other kind of proportional system is the relational one, in which the parts of the body are measured not by an arbitrary module but by their relation to the whole. When Vitruvius says that the size of the foot is one sixth of the whole body, he is not using the foot only as a module, but is referring to it as relational to the whole. In contrast to modular systems, which are projected upon the body, relational systems originate from the body itself. Therefore, relational systems close the body in on itself in a stronger manner than do modular systems. In them nothing is external to the body. In fact, by means of these systems the body is only mapped onto itself. Unlike modular systems, which are open, these systems have no loose ends. The whole is the border since all the parts are measured by it.

Thus, modular systems make a principle of individuation in the sense that through them an ideal is embodied in matter, but they do not create an individual that is closed upon itself and is separated from the rest of the world. Relational systems, on the other hand, individuate less by imposing a set of ideals on matter and more by closing the body on itself. This allows us to use it in a much more flexible manner, and to take into account the changes that occur in the body because of its dynamic nature without impairing its closure. Since relational systems originate from the organism itself and take into account its irregularities and its changes, while modular systems start from a standard "brick" that is multiplied mechanically, the latter is considered mechanical and the former organic.[3]

Rational and Irrational Systems. Another classification of proportional systems distinguishes between the systems based on rational relations and those based on irrational relations. According to the Vitalists, the main characteristic of the vital factor is that it is external to matter, but at the same time it is embodied in it and activates it. Although it is something external to matter, it is not imposed on it from outside (like a frame) but rather it controls the development and growth of the matter from within (like a driver of a car). Therefore, it is not necessary that relational systems should frame the object in clear physical borders. (Of course irrational systems are all relational since their definition includes the idea that there is no shared module to the parts.) Instead, they determine the autonomy of the object by virtue of the fact that its system of evolution is its own. While this condition can exist when one uses rational proportional systems, as in the system of musical harmony, it is more evident when the system is irrational, as in the proportions of the Golden Section [Scholfield 1958: 13].

In an irrational system, when we start from the whole and give it a rational number, then all the parts have irrational numbers as their measurement. Therefore, the size of the parts is defined only by their relation to the whole and do not have a definite numerical number of their own. Thus a strong conceptual dependency exists between the parts and the whole and between the different parts. This makes the system a closed one, so that all its parts depend on the formula of the series.

When one starts with the part as the definite number and not with the whole, the advantage of the irrational system, according to Matila Ghyka (1881-1965) is that it allows dynamic growth instead of mechanical addition. In contrast to the inorganic static world, in which the principle is that of minimum activity, the organic world is characterized by growth [Ghyka 1977: 85-86]. Growth is possible when the system is not totally closed, and thus it has some sort of lack which prevents the filling of the whole surface or volume. This lack is translated into potential energy because it "aspires" to be filled up. Since something is lacking in the symmetry of irrational systems, room is available for growth which is not simply an addition of parts but is the result of the object's own "aspiration" to complete the symmetry (which is never fulfilled in the irrational systems).[4] Since the origin of this growth is an internal energy and not an external addition, it detaches the object from its surroundings, with which it has no common measure. Thus, the fact that irrational systems are always open-ended is exactly what make them self sufficient. Here both the principle of growth and the principle of individuation unite in one principle of self sufficiency and autonomy. An example of this sort of growth is the spiral.

Although irrational systems create a sort of self-sufficiency, they nevertheless contain a lack that we can compare to the need of organisms for nutrition. The system has a certain closure because it cannot be divided without remainder, but that is also the reason that it contains a lack, i.e., an openness or, as it were, a "hole". This lack, or indefiniteness, is what makes it non-mechanical. This indefinite "hole" is the "place" where "life" can enter. In *Timaeus*, Plato ascribes the soul of the world to whole numbers and the body to irrational (geometric) numbers [Zeyl 2000: 17-23]. According to the idea presented here we might say that the "hole" in the body is what enables the soul – the vital force – to enter into it, thereby creating a living organism. Only when joined do the body and the vital force create a closed system.

Thus far we can see that modular systems of proportion create infinite and open systems – continuous with the world – while relational systems create closed and finite systems – separate from the world. Irrational systems enable the creation of a closed but infinite system and thus enable growth.

Proportional systems as equipotential

A fashionable system of proportion is that of "fractal geometry". The name "fractal" was coined by Benoit Mandelbrot for a geometric body in which a similar motif repeats itself at all scales. The simplest image of fractals is a tree in which every branch is divided in the same manner as the trunk itself is divided (see [Lauwerier 1991: xi, xii]). Fractals are characterized by intrinsic self-similarity and by an infinite ability to grow, both inwards and outwards.[5] Since the fractal resembles itself a all of its scales, it does not depend on actual size. In other words, the formula matches every size. In this regard it is no different from any other proportional system; when it is based on rational numbers one can even use every small part of it as a module.

As already mentioned, Driesch claims that the organism is an equipotential system. That is, in the early stages of the organism's development, every part of it has the potential to develop into either a whole organism or into any organ [Driesch 1908, 1: 132-159]. This is a sort of self-similarity – every part is similar to the whole – and in that respect the organism resembles fractal geometry. Fractal geometry is also used as a principle of growth. In that respect it is similar to the principle of growth in irrational numbers (which are actually a sort of fractal). In fractal geometry it does not matter whether we use rational or irrational systems, since the fractal itself is based on a formula of growth. Thus, even when only rational numbers are used, it does not give the feeling of addition but of growth which is directed from the inside. This growth occurs in a certain order but not necessarily a pre-conceived one.

Mathematical proportions as creators of harmony

Earlier, tubularia was mentioned as an example of equipotentiality, but it is also an example of a proportional harmony. As described above, no matter where one cuts the Tubularia it grows a proportional head. Since mathematical proportion is something formal and not causal, Driesch decided that we are dealing here not with a deterministic causal principle but with a formal principle which has its origin in something different from matter – some sort of vital factor. Analogously we can say about a building that has proportional correspondence between its parts that this correspondence derives from some sort of vital factor that is connected to neither the material structural factor of the building, nor to its functionality.

The more events we find that are harmonious with each other, the more difficult it is to assign this harmony to contingency. Rather, one must assume the existence of an external factor which is responsible for that harmony. Likewise, a proportional system that corresponds to multiple parts of the building gives us the feeling that it is not accidental, but contains an external "vital" factor responsible for it. That is, the existence of a proportional system points to something that coordinated the parts formally. We know, of course, that the architect is responsible for that correspondence, but because it appears not to be imposed on the building but rather intrinsic to it, we are inclined to assume that there is in the building itself something that is not its matter, and that this something is responsible for its harmony, as the soul is responsible for harmony in the body.

Thus, mathematical proportions should be something that are not identical to the building's matter (or body, or flesh) – that is, it should be external to it – but at the same time it should appear as if it is somehow intrinsic to it. According to Robin Evans, that might be the reason why in periods like the Renaissance mathematical principles of architecture were not explicit, but only revealed themselves to the beholder who knew how to interpret architecture and expose them. As an example, Evans contrasts the two different interpretations of the facade of St. Maria Novella by Wölfflin and Wittkower. In both, the geometric analysis is based on lines that are implicit in the building but are not expressed in any explicit way [Evans 1995: 248]. Thus, these lines of proportions are like a vital force that animates the building from within without being one with its body (matter). By being buried "deep" in the building, they act like its soul.

In order to enliven the building, the mathematical ideal should not only be buried in it; it should also somehow not totally overlap its physical embodiment. That is, there should be something which is not totally accurate in the application of mathematics to the actual building. For example, as Rhys Carpenter points out, there are always some inaccuracies in Greek temples. These, according to him, are not the outcome of inaccuracies of execution, but are the result of a deliberate intention to enliven the mathematical ideal [Carpenter 1970: 14-15]. In the same vein Fernando Perez Oyarzun says that in architecture, like in the organism, the ideal principle and its actual embodiment do not fully overlap each other. As an example he cites Leonardo's version of the Vitruvian man, in which there is no complete overlap between the square and the circle [Perez Oyarzun 1997: 223-224].

Notes

1. Vitruvius himself is not clear about this. In bk. IV, chap. i, when he talks about the origin of the orders he tends toward the first sense of the similarity but in other places, as in bk. I, chap. ii, or here in bk. III, chap. i, he only says that there are proportions in both buildings and human beings without saying that they are the same proportions.

2. See [Panofsky 1983: 83]. Scholfield mentions both systems in relation to Vitruvius [1958: 2]. See also [Arnheim 1966: 111-112].

3. See [Panofsky 1983: 93, 102]. He makes a distinction between the use of module in ancient Egypt, which he calls constructive, and the use of the module in Greece, which he calls anthropomorphic.

4. The word symmetry is used here in the Vitruvian sense as Ghyka uses it, referring to the proportional relations that create a relation between the parts and the whole; see [Ghyka 1977: 5].

5. So that the growth will have self-similarity yet avoid monotony, transformations occur in the similarity: rotation, central enlargement, circular enlargement, reflection, etc. Also, it is common to insert some accidental factors in the fractal transformation by relating to the similarity as statistical and not total. In other words, instead of having the same qualities each part of the fractal system has the same statistical qualities; see [Lauwerier 1991: 84, 104].

References

ARNHEIM, Rudolf. 1966. *Toward a Psychology of Art*. Berkeley: University of California Press.

DRIESCH, Hans. 1908. *The Science and Philosophy of the Organism*, 2 vols. London: Adam and Charles Bloch.

VITRUVIUS. 1960. *The Ten Books on Architecture*. Morris Hickey Morgan, trans. New York: Dover.

PANOFSKY, Erwin. 1983. *Meaning in the Visual Arts*. Harmondsworth, UK: Penguin Books.

SCHOLFIELD, P. H. 1958. *The Theory of Proportion in Architecture*. Cambridge, UK: Cambridge University Press.

GHYKA, Matila. 1977. *The Geometry of Art and Life*. New York: Dover Publications.

ZEYL, Donald J., trans. 2000. *Plato's Timaeus*. Indianapolis: Hackett Publishing Company.
LAUWERIER, Hans. 1991. *Fractals*. Princeton: Princeton University Press.
EVANS, Robin. 1995. *The Projective Cast*. Cambridge, MA: MIT Press.
CARPENTER, Rhys. 1970. *The Architects of the Parthenon*, Harmondsworth, UK: Penguin Books.
PEREZ OYARZUN, Fernando. 1997. The Mirror and the Cloak. In: *Anybody*, C. Cynthia Davidson, ed. Cambridge, MA: MIT Press.

About the author

Anat David Artman is an architect who has taught in the department of architecture at Bazalel Academy of Art since 1992. In addition she studied philosophy in the Hebrew University and completed her Ph.D. in philosophy in 2006.

Christopher Stone

AFEC
University of Derby
Kedleston Road
Derby DE22 1GB
UNITED KINGDOM
c.p.stone@derby.ac.uk

Keywords: algebra, algorithms,
circles, design theory, fractals,
geometry, infinity, iteration,
lines, mappings, morphology,
ratio, rectangles, shape
grammars, topology,
transformations

Didactics

The Use of Linear Fractional Transformations to Produce Building Plans

Abstract. Linear fractional transformations are mappings on the complex plane that can be used to form building plans. Linking topological and geometrical considerations to meet certain criteria is one way of generating plans. This present study seeks any mathematical structures underlying basic plan forms which would link topological and geometrical maps. It concludes with an algorithm for plan generation.

"The plan is the generator" said Le Corbusier; but how are plans generated? We usually start with a topological map (bubble diagram) and morph this into a geometrical map (plan). I was interested to find a mathematical structure underlying this process which was, perhaps, more elegant than the "chopping block" techniques of most shape grammar methods.

Linear fractional transformations (LFTs) are mappings on the complex plane and have been used to produce fractal patterns. I have used the same transformations to generate rectangular configurations to form building plans.

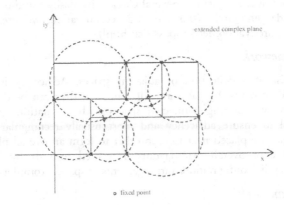

Fig. 1.

A particular rectangle can be generated by inscribing it in a particular circle starting at a specific point on the circumference. LFTs map circles to circles and if the circles overlap this will generate further adjacent rectangles which form architectonic patterns. To ensure that circles overlap the transformations have to be elliptic. This ensures that there are two fixed points and the position of these determines the way the layout progresses. By varying the parameters *a, b, c* and *d* a range of plan types is produced using mathematical software.

This is a first small step in a new method of plan generation.

Introduction

The main problem of space layout planning is linking topological considerations with geometrical ones to meet certain criteria [Jo and Gero 1998]. Topological considerations of connectedness and adjacency of spaces can be represented by a graph comprising a set of vertices and edges, familiar to architects as a bubble diagram. To transform this into a building plan involves turning the vertices into geometrical shapes (usually rectangles) while considering spatial, access and aesthetic requirements. Formal enumeration of these transformations is fraught with problems, not least the 'combinatorial explosion' of numbers of possible solutions [Steadman 1989: 112]. In practice architects and designers make tentative arrangements of shapes sometimes using circular or elliptical forms to approximate the rectangles before the layout is firmed up. Many approximations may be made and at each stage a selection is made based on intuitive skill.

I wanted to investigate the possibility of finding any mathematical structures underlying basic plan forms which would link topological and geometrical maps.

There is currently much mathematical interest in packings of circles into geometric shapes [Stephenson 2005] and the exploration of soap bubbles [Oprea 2000: 43-50], both of which make extensive use of mathematics in the complex plane. Complex analysis has also been used to map the surface of the brain, explain Escher's drawing "Hand with Reflecting Sphere", analyse Celtic art and generate fractal patterns (see [Hurdal 2003]; [Blair 2000: iv]; [Ellacott 2000] and [Mumford et al, 2002]). With such fertile links with geometry I wanted to see if this area of maths had applications in architectural design.

A final inspiration was the picture "Several Circles" by Vasily Kandinsky which consists of a pattern of overlapping circles (this is reproduced on various websites; see for instance http://guggenheim.stores.yahoo.net/kanpossevcir.html).

Architectural framework

Buildings consist of a collection of adjacent spaces. Adjacency is the key to the production of building plans. If spaces are adjacent then they can be directly connected. Plans are built up by grouping connected spaces with circulation spaces. Spaces are generally polygonal to ensure adjacency and most usually rectangular or made up of rectangles. Rectangles are placed next to each other to form an overall plan layout. There are several categories of overall plan layouts. I wanted to start by looking at linear configurations and cyclic configurations (arrangements of spaces around a common space).

Mathematical framework

A rectangle of any size, proportion and position can be generated by inscribing horizontal and vertical lines from a point on a circle in the Cartesian plane. An adjacent rectangle can be generated by inscribing it in another circle that overlaps the first and passes through at least one vertex of the first rectangle. The vertex chosen and the size of the second circle determine the size, proportion and configuration of the second rectangle. A plan can be thus generated by a series of overlapping circles. The process can be pictured as a circle flipping over about two points on its perimeter while changing size. At each flip a rectangle will be left.

On the complex plane points are represented by numbers composed of two parts; a real number that appears on the conventional horizontal number line and an imaginary number[1] that is plotted on a vertical scale. The imaginary number is a multiple of $\sqrt{(-1)}$ (the square root of minus one denoted as i). A complex number $(x + iy)$ is the sum of a real number and an imaginary one and is both a number and a point on a plane. Points on the plane can be moved around by adding, subtracting, multiplying or dividing these numbers. This property means that geometrical operations are generally simpler compared with the matrix operations required on a real Cartesian plane.

Conformal mapping on the complex plane preserve certain geometrical properties. Linear fractional transformations (or Moebius transformations) map circles to circles [Needham 2000: 148].[2] LTRs are of the form $w = \dfrac{az+b}{cz+d}$ where a, b, c and d are complex parameters and z and w are complex variables. A circle on the z plane can be transformed to a circle on the w plane and these planes can be displayed together. The extended complex plane includes the point at infinity and part of the underlying structure of these operations involves the poles of the transformations. There are two poles. z_∞ is the point that is taken to the point at infinity and Z_∞ is the point that the point at infinity is transformed back to [Needham 2000: 170]. They are defined by:

$$z_\infty = -\frac{d}{c} \qquad \text{and} \qquad Z_\infty = \frac{a}{c}$$

The poles are included for completeness and have been plotted and tabulated with other results.

There are four categories of LFTs depending on the values of a, b, c and d [Needham 2000: 150]. In two cases the circles do not overlap (hyperbolic and loxodromic). In one case the circles touch at one point (parabolic) and in one case the circles overlap (elliptic). A transformation is elliptical iff $(a+d)$ is real and $|a+d| < 2$ [*Complex Analysis* 1995: 23]. This can be written as $0 \leq (a + d)^2 < 4$. This condition applies to normalised transformations. For a transformation to be normalised a, b, c and d have to be multiplied by $\dfrac{\pm 1}{\sqrt{ad-bc}}$. To ensure an LFT is elliptical it would have to be normalised and the condition checked but the LFT does not have to be used in its normalised form [Schwerdtfeger 1979: 41].

If two rectangles are adjacent their circumscribed circles will overlap. The LFT transforming one circle to the other can be found from:

$$\frac{(z-\alpha)(\beta-\gamma)}{(z-\gamma)(\beta-\alpha)} = \frac{(w-\alpha')(\beta'-\gamma')}{(w-\gamma')(\beta'-\alpha')}$$ this maps $\alpha \to \alpha'$, $\beta \to \beta'$ and $\gamma \to \gamma'$ where these are any three points on the first circle transformed to three points on the other.

A circle is defined uniquely by three points and the LFT of the three points will uniquely transform one circle to the other.

Investigations

I have arranged rectangles in six linear configurations (Transformations A-F) and three cyclic configurations (Transformations G, H and J). Transformations A and B show lines of squares and C and D show 2x1 rectangles. Points that are fixed and that are to be transformed were selected (three in total) and substituted in the above formula and the parameter values were generated using *DERIVE*. The direction of transformations is shown and the path of the z pole is indicated as are the overlapping circles and fixed points. In this case both fixed points are also vertices of the rectangles:

Transformations A-B are shown in fig. 2. Transformations C-D are shown in fig. 3. Results are shown in **Table 1 (A-D)**.

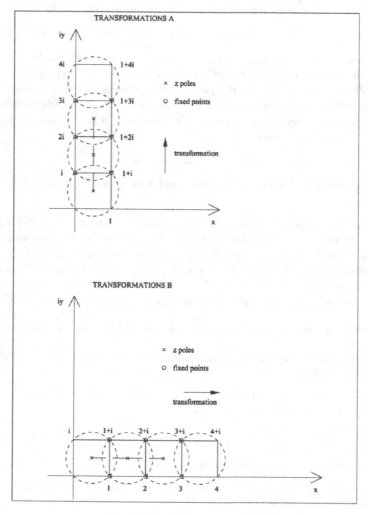

Fig. 1. Transformations A and B

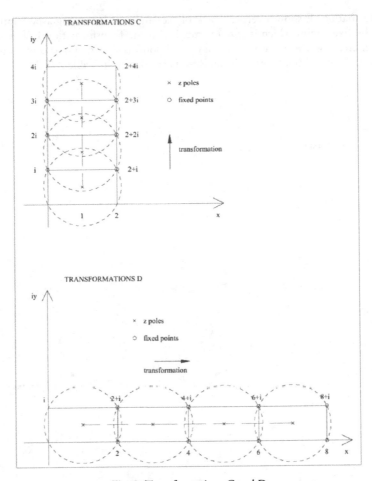

Fig. 2. Transformations C and D

	a	b	c	d	Pole z_∞	Pole Z_∞	Fixed Points
A ↑	$-(2n+)+i$	$2n-(2n^2)i$	$2i$	$(2n-1)-i$	$\frac{1}{2}+\frac{(2n-1)i}{2}$	$-\frac{1}{2}+\frac{(2n+1)i}{2}$	$n,1+ni$
B →	$-1+(2n+1)i$	$2n-(2n^2)i$	$2i$	$1-(2n-1)i$	$\frac{(2n-1)}{2}+\frac{i}{2}$	$\frac{(2n+1)}{2}+\frac{i}{2}$	$n,n+i$
C ↑	$-(2n+1)+2i$	$4n+(2n^2)i$	$2i$	$(2n-1)-2i$	$1+\frac{(2n-1)i}{2}$	$-1+\frac{(2n+1)i}{2}$	$ni,2+ni$
D →	$-1+(4n+2)i$	$4n-(8n^2)i$	$2i$	$1-(4n-2)i$	$(2n-1)+\frac{i}{2}$	$(2n+1)+\frac{i}{2}$	$2n,2n+i$

Table 1 (A-D). Results are summarised by creating formulae where n represents the order number of the transformation starting at 1

Transformations E show a vertical stack of rectangles of increasing aspect ratio. In this case only the fixed point which is also a vertex is shown. Transformation F show a stack of squares of increasing size and again only one fixed point is shown. Transformations E and F are shown in fig. 4 and fig. 5 respectively. Results are shown in **Table 2 (E,F)**.

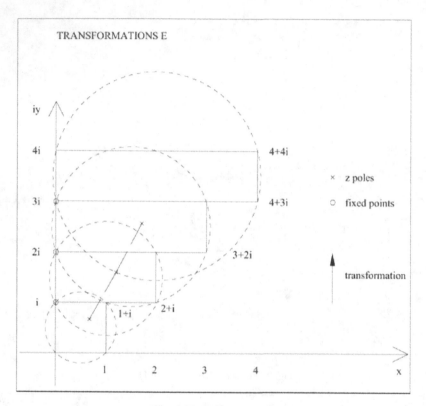

Fig. 4. Transformations E

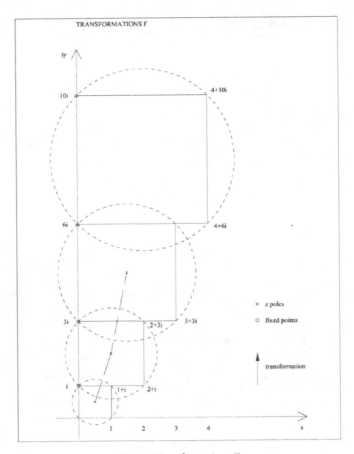

Fig. 5. Transformations F

	a	**b**	**c**	**d**
E ↑	$(2n^2 + 2n + 1) + (n^2 + n)i$	$(2n^2 + 2n + 1) + (n^2 + n)i$	$(2n+1)i$	$2n^2 - (n^2 + n)i$
F ↑	$-(n+1)^2 + (n+1)i$	$(n^3 + \dfrac{3n^2}{2} + \dfrac{n}{2}) + (\dfrac{4n^3}{3} + \dfrac{7n^2}{2} - \dfrac{11n}{6})i$	$2i$	$n^2 - ni$

	Pole z_∞	Pole Z_∞	Fixed Point
E ↑	$\dfrac{n(n+1)}{2n+1} + \dfrac{2n^2 i}{2n+1}$	$\dfrac{n(n+1)}{2n+1} + \dfrac{(2n^2 + 2n + 1)i}{2n+1}$	ni
F ↑	$\dfrac{n}{2} + \dfrac{n^2 i}{2}$	$\dfrac{n+1}{2} + \dfrac{(n+1)^2 i}{2}$	$\dfrac{n(n+1)i}{2}$

Table 2 (E,F). Results are summarised by creating formulae where *n* represents the number of the transformation

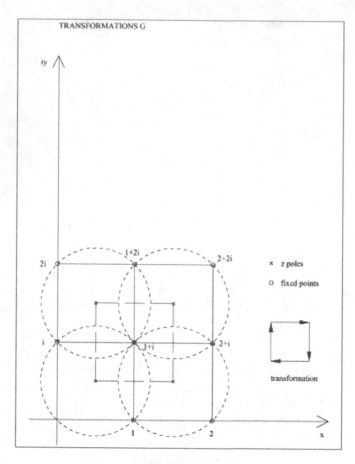

Fig. 6. Transformations G

Transformations G show a cyclic arrangement of squares and both fixed points are shown (fig. 6). Results are shown in **Table 3 (G)**.

	a	**b**	**c**	**d**	Pole z_∞	Pole Z_∞	Fixed Points
↑	$-3+i$	$2+2i$	$2i$	$1-i$	$\frac{1}{2}+\frac{i}{2}$	$\frac{1}{2}+\frac{3i}{2}$	$i, 1+i$
→	$-3+3i$	$6+2i$	$2i$	$3-i$	$\frac{1}{2}+\frac{3i}{2}$	$\frac{3}{2}+\frac{3i}{2}$	$1+2i, 1+i$
↓	$-1+3i$	$6-2i$	$2i$	$3-3i$	$\frac{3}{2}+\frac{3i}{2}$	$\frac{3}{2}+\frac{i}{2}$	$2+i, 1+i$
←	$-1+i$	$2-2i$	$2i$	$1-3i$	$\frac{3}{2}+\frac{i}{2}$	$\frac{1}{2}+\frac{i}{2}$	$1, 1+i$

Table 3 (G). (Results cannot be easily summarised by creating formulae)

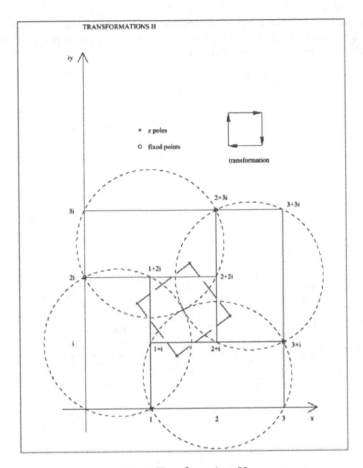

Fig. 7. Transformations H

Transformations H show a cyclic arrangement of 2x1 rectangles around a square. One fixed point per transformation is shown (fig. 7). Results are shown in **Table 4 (H)**.

	a	b	c	d	Pole z_∞	Pole Z_∞	Fixed Point
↑	$-14+2i$	$12+24i$	$5i$	$8-4i$	$\dfrac{4}{5}+\dfrac{8i}{5}$	$\dfrac{2}{5}+\dfrac{14i}{5}$	$2i$
→	$-13+14i$	$54+3i$	$5i$	$11-8i$	$\dfrac{8}{5}+\dfrac{11i}{5}$	$\dfrac{14}{5}+\dfrac{13i}{5}$	$2+3i$
↓	$-1+13i$	$18-24i$	$5i$	$7-11i$	$\dfrac{11}{5}+\dfrac{7i}{5}$	$\dfrac{13}{5}+\dfrac{i}{5}$	$3+i$
←	$-2+i$	$6-3i$	$5i$	$4-7i$	$\dfrac{7}{5}+\dfrac{4i}{5}$	$\dfrac{1}{5}+\dfrac{2i}{5}$	1

Table 4 (H). (Results cannot be easily summarised by creating formulae)

Transformations J show a cyclic arrangement of squares and rectangles around a rectangular and looks more like a building plan. One fixed point per transformation is shown (fig 8). Results are shown in **Table 5 (J)**.

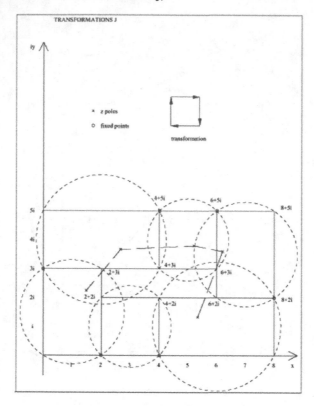

Fig. 8. Transformations J

	a	b	c	d	Pole z_∞	Pole Z_∞	Fixed Point
↑	$-18+4i$	$30+45i$	$4i$	$9-6i$	$\dfrac{3}{2}+\dfrac{9i}{4}$	$1+\dfrac{9i}{2}$	$3i$
→	$-13+16i$	$96-3i$	$3i$	$11-8i$	$\dfrac{8}{3}+\dfrac{11i}{3}$	$\dfrac{16}{3}+\dfrac{13i}{3}$	$4+5i$
→	$-19+36i$	$238+127i$	$5i$	$19-26i$	$\dfrac{26}{5}+\dfrac{19i}{5}$	$\dfrac{36}{5}+\dfrac{19i}{5}$	$6+5i$
↓	$-4+36i$	$146-198i$	$5i$	$18-31i$	$\dfrac{31}{5}+\dfrac{18i}{5}$	$\dfrac{36}{5}+\dfrac{4i}{5}$	$8+2i$
←	$-2+8i$	$24+48i$	$3i$	$4-16i$	$\dfrac{16}{3}+\dfrac{4i}{3}$	$\dfrac{8}{3}+\dfrac{2i}{3}$	4

Table 5 (J). (Results cannot be easily summarised by creating formulae)

Observations

(i)   There is a consistency about parameter *c* which only appears to vary with the proportions of successive rectangles;

(ii)   There is some interchange of real and imaginary values with opposite signs for *a* and *d* for successive transformations;

(iii) Clearly the lack of pattern in the parameters grows as the layouts become more complicated particularly with parameter *b* (in this case it may not be a problem);[3]

(iv)   The poles shown follow a less consistent path as the layout becomes more complicated.

Conclusions

There is sufficient information to enable a selection of parameters to be made to feed into an algorithm to generate plan forms something like this:

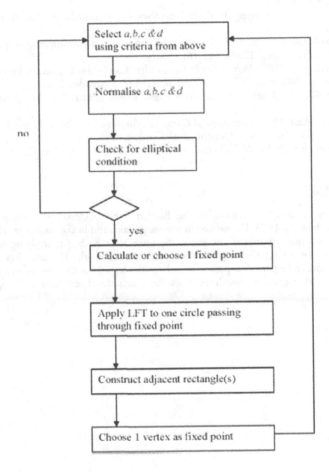

Notes

1. A slightly pejorative term reflecting the awkward history of this branch of mathematics. When devised it was clear that these numbers could not be used for counting or measuring so where not considered proper numbers. They are now used in quantum physics to do just that and complex analysis is regarded as a basic skill in higher mathematics)
2. Strictly straight lines through the point at infinity appear as straight lines but are regarded as generalised circles. Angles of intersection of circles are preserved)
3. The fixed points are at $\xi_\pm = \dfrac{(a-d) \pm \sqrt{(a+d)^2 - 4}}{2c}$ and so are independent of b).

References

BLAIR, David E. 2000. *Inversion Theory and Conformal Mapping.* Rhode Island: The American Mathematical Society.

COMPLEX ANALYSIS. 1995. *Complex Analysis, Unit D1, Conformal Mappings.* Milton Keynes: The Open University, 1995.

ELLACOTT, Stephen. 2000. The Arithmetical Structure of Celtic Key Patterns. *Mathematics Today* (October 2000): 142-145.

HURDAL, Monica A. 2003. Imaging the Brain. http://www.math.fsu.edu/~mhurdal/. Last updated on 21.07.2007.

JO, Jun H. John S. Gero. 1998. Space Layout Planning using an Evolutionary Approach. *Artificial Intelligence in Engineering* **12** (3): 149-162.

MUMFORD, David, et al. 2002. *Indra's Pearls.* Cambridge: Cambridge University Press.

NEEDHAM, Tristan. 2000. *Visual Complex Analysis.* Oxford: Clarendon Press.

OPREA, John. 2000. *The Mathematics of Soap Films.* Rhode Island: The American Mathematical Society.

SCHWERDTFEGER, Hans. 1979. *Geometry of Complex Numbers.* New York: Dover Publications.

STEADMAN, J.P. 1989. *Architectural Morphology.* London: Pion.

STEPHENSON, Kenneth. 2005. *Introduction to Circle Packing.* Cambridge: Cambridge University Press.

About the author

Christopher Stone studied architecture at the Bartlett School, University College London and qualified as an architect in 1973. He worked in private practice and in the transport sector until 1994 when he began working for myself. At about the same time he began studying for a degree in mathematics and combining teaching with my architectural work. He now has an established architectural practice and a part-time post as a mathematics lecturer at the University of Derby, where he runs an entry level programme. See: http://www.derby.ac.uk/staff-search/mr-christopher-stone.

Rachel Fletcher

113 Division St.
Great Barrington, MA
01230 USA
rfltech@bcn.net

Keywords: descriptive geometry, diagonal, dynamic symmetry, incommensurate values, root rectangles

Geometer's Angle

Dynamic Root Rectangles Part Three: Root-Three Rectangles, Palladian Applications

Abstract. "Dynamic symmetry" is the name given by Jay Hambidge to describe a system of incommensurable ratios for proportioning areas within design compositions. In Parts One and Two of a continuing series, we surveyed the elements of root-two, -three, -four, and –five rectangular systems and, using the root-two rectangle, explored diagonals, reciprocals, complementary areas, and other techniques for composing dynamic space plans. In Part Three, we apply these techniques to the root-three rectangle and consider architectural plans by Andrea Palladio.

Introduction

"Dynamic symmetry," is the name given by Jay Hambidge to describe a system of proportion for rectangles of square root proportions, where the incommensurable ratio that measures the rectangle as a whole replicates through endless divisions.[1] Hambidge traces this method for proportioning areas from origins in ancient Egypt to sixth or seventh century BC Greece, where it subsequently developed in Euclidean geometry and for three hundred years produced some of the finest art of the Classical period [Hambidge 1920, 7-8].

In Hambidge's method, the proportion of the enclosing rectangle governs the spatial elements within. One example is the root-three rectangle, whose ratio of $1:\sqrt{3}$ cannot be expressed in finite numbers, but is precisely the relationship of the half-side of an equilateral triangle to its altitude. The root-three rectangle can be formed from the two sides of a 30°-60°-90° triangle. In this article we explore its dynamic properties and, in celebration of Andrea Palladio's quincentenary, consider the potential for spatial composition in two of the architect's designs.

Review: The Root-Three Rectangle, its Diagonal and Reciprocal

How to draw a root-three rectangle

- Draw a horizontal baseline AB equal in length to one unit.
- From point A, draw an indefinite line perpendicular to line AB that is slightly longer in length.
- Place the compass point at A. Draw a quarter-arc of radius AB that intersects line AB at point B and the indefinite line at point C.
- Place the compass point at B. Draw a quarter-arc (or one slightly longer) of the same radius, as shown.
- Place the compass point at C. Draw a quarter-arc (or one slightly longer) of the same radius, as shown.
- Locate point D, where the two quarter-arcs (taken from points B and C) intersect.

- Place the compass point at D. Draw a quarter-arc of the same radius that intersects the indefinite vertical line at point C and the line AB at point B (fig. 1).
- Connect points D, B, A and C.

The result is a square (DBAC) of side 1 (fig. 2).

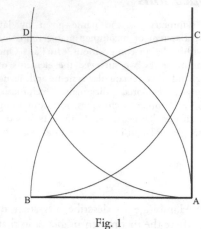
Fig. 1

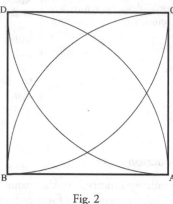
Fig. 2

- Draw the diagonal BC through the square (DBAC).

The side (DB) and the diagonal (BC) are in the ratio 1:$\sqrt{2}$.[2]

- Place the compass point at B. Draw an arc of radius BC that intersects the extension of line BA at point E (fig. 3).

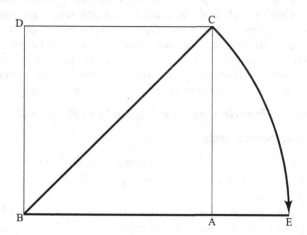
Fig. 3

- From point E, draw a line perpendicular to line EB that intersects the extension of line DC at point F.
- Connect points D, B, E and F.

The result is a root-two rectangle (DBEF) with short and long sides in the ratio 1:√2 (fig. 4).

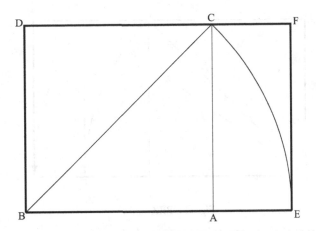

Fig. 4. DB:BE :: 1:√2

- Draw the diagonal BF through the root-two rectangle (DBEF).

The side (DB) and the diagonal (BF) are in the ratio 1:√3.

- Place the compass point at B. Draw an arc of radius BF that intersects the extension of line BE at point G (fig. 5).

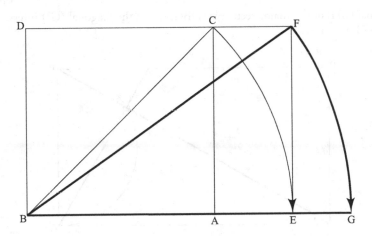

Fig. 5

- From point G, draw a line perpendicular to line GB that intersects the extension of line DF at point H.
- Connect points D, B, G and H.

The result is a root-three rectangle (DBGH) with short and long sides in the ratio 1:√3 (fig. 6).

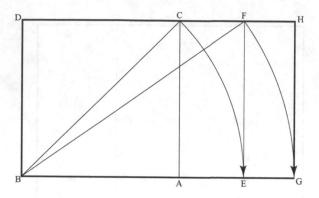

Fig. 6. DB:BG :: 1:√3

How to divide a root-three rectangle into reciprocals

Root rectangles produce reciprocals of the same proportion when the diagonal of the major rectangle and the diagonal of its reciprocal intersect at right angles.[3]

- Locate the diagonal (BH) of the rectangle DBGH.
- Locate the line GH. Draw a semi-circle that intersects the diagonal BH at point O, as shown.
- From point G, draw a line through point O that intersects line HD at point I.

The diagonal (BH) of the major rectangle (DBGH) and the diagonal (GI) intersect at right angles (fig. 7).

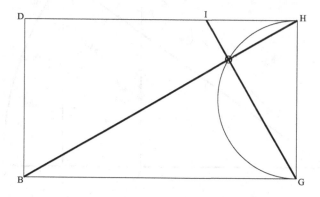

Fig. 7. GI:BH :: 1:√3

- From point I, draw a line perpendicular to line HD that intersects line BG at point J.
- Connect points J, G, H and I.

The result is a smaller root-three rectangle (JGHI) with short and long sides of $1/\sqrt{3}$ and 1 ($\sqrt{3}/3$:1 or 0.5773…:1). Rectangle JGHI is the reciprocal of the whole rectangle DBGH. The major 1:$\sqrt{3}$ rectangle DBGH divides into three reciprocals (JGHI, LJIK and BLKD) that are proportionally smaller in the ratio 1:$\sqrt{3}$. The area of each reciprocal is one-third the area of the whole (fig. 8).

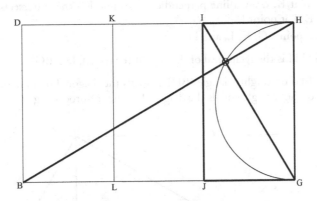

Fig. 8. IH:HG :: HG:GB :: 1:$\sqrt{3}$

The diagonal (BH) of the major rectangle (DBGH) and the diagonal (GI) of the reciprocal (JGHI) locate endless divisions in continual proportion. A root-three rectangle of any size divides in equal parts into three reciprocals in the ratio 1:$\sqrt{3}$. Each reciprocal divides in similar fashion.

As the process continues, the side lengths of successively larger rectangles form a perfect geometric progression (1 ,$\sqrt{3}$, 3, 3$\sqrt{3}$…). The side lengths of successively smaller rectangles decrease in the ratio 1:1/$\sqrt{3}$ toward a fixed point of origin known as the pole or eye (point O). (See figure 9.)

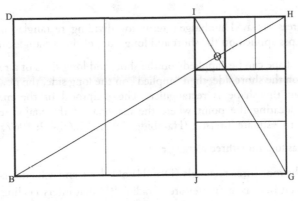

Fig. 9

- Locate the root-three rectangle (DBGH) of sides 1 and $\sqrt{3}$, and its diagonal BH.

- From point H, draw an indefinite line perpendicular to the diagonal HB.
- Extend the line BG until it intersects the indefinite line at point K, as shown.
- Connect points B, H, and K.

The result is a right triangle (BHK).

- From point K, draw a line perpendicular to line KB that intersects the extension of line DH at point L.
- Connect points G, K, L, and H.

The rectangle GKLH is the reciprocal of the root-three rectangle DBGH.

If the long side (BH) of a right triangle (BHK) equals the diagonal of a major rectangle, the short side (KH) of the triangle equals the diagonal of the reciprocal (fig. 10).

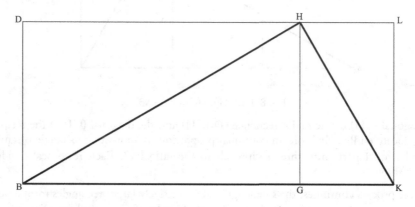

Fig. 10.   GK:KL :: DB:BG :: 1:√3   KH:HB :: 1:√3

Review: Application of Areas

Definition:

Application of areas is Jay Hambidge's term for dividing rectangles into proportional figures, where shapes applied on the short and long sides of the rectangle are equal in area.

Any quadrilateral shape can be "applied" on the short and long sides of a rectangle. When a square is applied on the short side, then "applied" on the long side, the new area is the same as the square, but the shape is rectangular. The reciprocal of the major rectangle is accomplished by locating the point where the diagonal of the major rectangle and the inside edge of the true square intersect (Hambidge 1960, 35, 60–72; 1967, 28–29).[4]

How to apply areas to a root-three rectangle

- Locate the root-three rectangle (DBGH) of sides 1 and √3, and its diagonal BH.
- From point B, draw a quarter-arc of radius BD that intersects line BG at point M.
- From point M, draw a line perpendicular to line BG that intersects line HD at point N.

The result is a square (DBMN) "applied" on the short side (DB) of the root-three rectangle (DBGH).

- Locate the diagonal BH of the root-three rectangle (DBGH).
- Locate point P where the diagonal (BH) intersects line MN.
- Draw a line through point P that is perpendicular to line MN and intersects line DB at point Q and line GH at point R.
- Locate the rectangle QBMP.

The rectangle (QBMP) is the reciprocal of the root-three rectangle DBGH. Line BP is the diagonal of the reciprocal (QBMP). (See figure 11.)

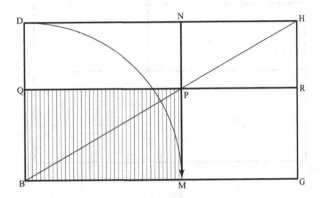

Fig. 11. QB:BM :: DB:BG :: 1:√3   BP:BH :: 1:√3

- Locate the square (DBMN) and the rectangle BGRQ.

The area of the rectangle (BGRQ) and the area of the square (DBMN) are equal. In Hambidge's system, the "square" is applied on the short and long sides of the root-three rectangle (fig. 12).

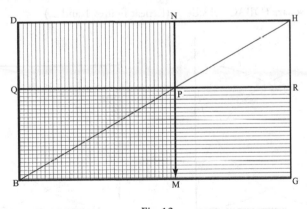

Fig. 12

Locate the reciprocal QBMP of the major root-three rectangle DBGH.

- Locate the rectangles DQPN and MGRP.

Rectangle DQPN is the complement of the reciprocal QBMP.[5]

The areas of rectangles DQPN and MGRP are equal.

- Locate rectangles QBMP and NPRH.

The rectangles QBMP and NPRH share the same diagonal and are similar (fig. 13).[6]

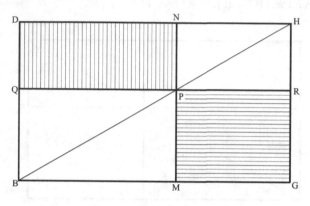

Fig. 13

The Root-Three Rectangle and Variations on Area Themes

The application of areas permits the division of root rectangles into harmonious compositions whose interior elements reflect the proportions of the overall figure. Let us consider the root-three rectangle and related figures.[7]

How to divide a square into root-three proportional areas

- Draw a square (DBAC) of side 1. (Repeat figures 1 and 2.)

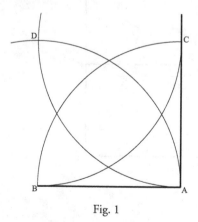

Fig. 1

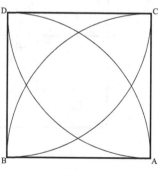

Fig. 2

- Locate points E, F, G, and H where the four quarter arcs intersect, as shown.
- From point B, draw a line through point H that intersects line AC at point I.
- From point A, draw a line through point F that intersects line DB at point J.
- Connect points I and J.

The result is a root-three rectangle (JBAI) of sides 1/√3 and 1 (0.5773… and 1). (See figure 14.)

- From point C, draw a line through point F that intersects line DB at point K.
- From point D, draw a line through point H that intersects line AC at point L.
- Connect points K and L.

The result is a root-three rectangle (LCDK) of sides 1/√3 and 1 (0.5773… and 1). (See figure 15.)

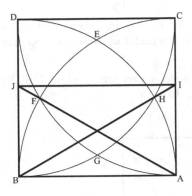

Fig. 14   Fig. 15

- From point D, draw a line through point G that intersects line BA at point M.
- From point B, draw a line through point E that intersects line CD at point N.
- Connect points M and N.

The result is a root-three rectangle (NDBM) of sides 1/√3 and 1 (0.5773… and 1). (See figure 16.)

- From point A, draw a line through point E that intersects line CD at point O.
- From point C, draw a line through point G that intersects line BA at point P.
- Connect points O and P.

The result is a root-three rectangle (PACO) of sides 1/√3 and 1 (0.5773… and 1). (See figure 17.)

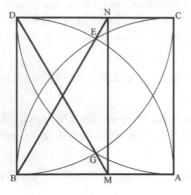
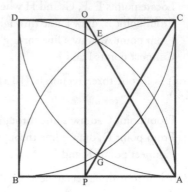

Fig. 16   Fig. 17

When root-three rectangles (JBAI, LCDK, NDBM and PACO) are drawn on all four sides of a square, the result is a composition that contains five center squares, four larger corner squares, and four root-three rectangles.

Each root-three rectangle (such as JBAI) has a complementary area (such as DJIC) that contains three squares and a root-three rectangle (fig. 18).

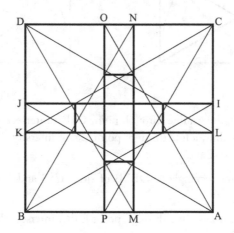

Fig. 18

Alternative Divisions and Design Applications

Option 1

- Locate the square (DBAC) of side 1.
- Locate the diagonals (DM, DL, BI, BN, AO, AJ, CK, and CP) of the four root-three rectangles, as shown (fig. 19).
- Connect points I and J.

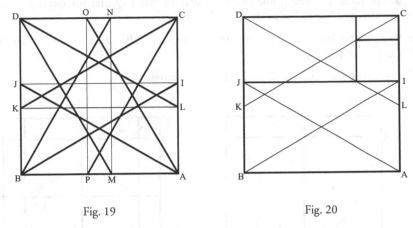

Fig. 19   Fig. 20

The result is a root-three rectangle (JBAI) of sides 1/√3 and 1 (0.5773... and 1).

When a root-three rectangle (JBAI) is drawn on the side (BA) of a square, the complementary area (DJIC) contains two root-three rectangles and a square (fig. 20).

Figure 21 repeats the scheme, with some modifications, on all four sides of the square.

Figure 22 presents a more complex design application.

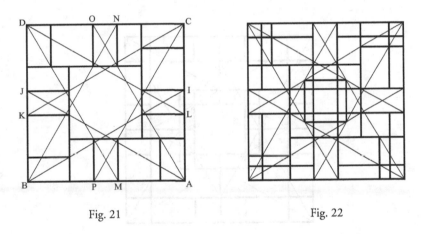

Fig. 21   Fig. 22

Option 2

- Locate the square (DBAC) of side 1.
- Locate the four quarter-arcs taken from points D. B, A, and C.
- Locate point E where the quarter-arcs taken from points B and A intersect.
- Locate point G where the quarter-arcs taken from points C and D intersect.
- From point E draw a line through point G that intersects line BA at point Q.

- From point E, draw a line perpendicular to line EQ that intersects line DB at point R and line AC at point S.

The result is a pair of two adjacent root-three rectangles (BQER and QASE) of sides ½ and √3/2 (0.5… and .8660) (fig. 23).

Figure 24 repeats the scheme on all four sides of the square.

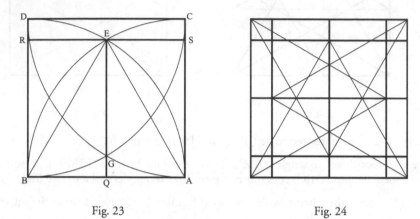

Fig. 23   Fig. 24

Figure 25 presents a more complex design application.

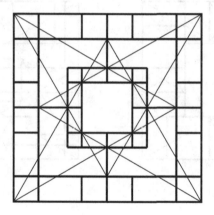

Fig. 25

How to divide a root-three rectangle into segments that progress in the ratio 1:√3

- Draw a root-three rectangle (DBGH) of sides 1 and √3. (Repeat figures 1, 2, 3, 4, 5 and 6.)

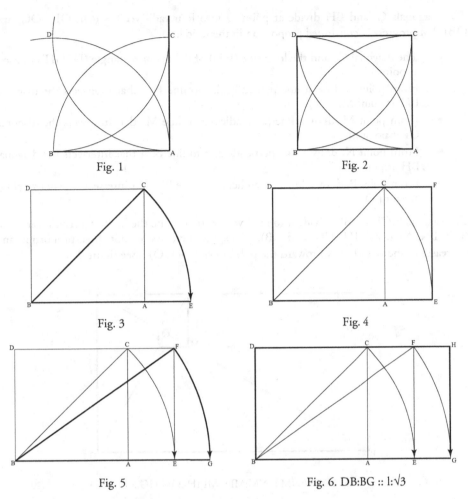

Fig. 1

Fig. 2

Fig. 3

Fig. 4

Fig. 5

Fig. 6. DB:BG :: 1:√3

- Locate the diagonal (BH) of the major rectangle (DBGH) and the diagonal GI, which intersect at right angles and are in the ratio 1:√3. (Repeat figure 7.)

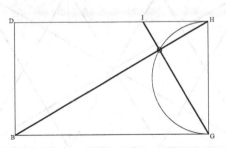

Fig 7 GI·BH :: 1:√3   OI:OH :: OH:OG :: OG:OB :: 1:√3

The diagonals GI and BH divide at point O into four radii vectors (OI, OH, OG, and OB) that progress in continued proportion in the ratio 1:√3.[8]

- The diagonal GI and the long side (HD) of the major rectangle (DBGH) intersect at point I.
- From point I, draw a line perpendicular to line HD that intersects the diagonal BH at point M.
- From point M, draw a line perpendicular to line MI that intersects the diagonal GI at point N.
- From point N, draw a line perpendicular to line NM that intersects the diagonal BH at point P.
- From point P, draw a line perpendicular to line PN that intersects the diagonal GI at point Q.

The diagonals GI and BH divide into radii vectors that locate the lines of a rectilinear spiral (QP, PN, NM, MI, IH, HG, and GB). The spiral increases in root-three proportion and decreases in the ratio 1: 1/√3 toward the pole or eye (point O). (See figure 26.)

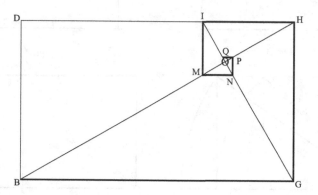

Fig. 26. PN:NM :: NM:MI :: MI:IH :: IH:HG :: 1:√3

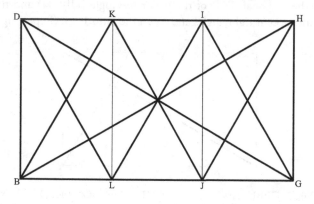

Fig. 27

- Draw the diagonals BH and DG of the root-three rectangle DBGH, the diagonals GI and JH of the reciprocal JGHI, the diagonals JK and LI of the reciprocal LJIK, and the diagonals LD and BK of the reciprocal BLKD (fig. 27).
- Use the diagonals to repeat the equiangular spiral three times, as shown (fig. 28).

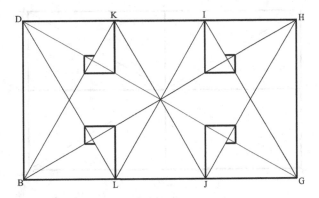

Fig. 28

- Locate points R and S, as shown.
- Connects points L, J, R, I, K, and S.

The result is a hexagon.

- Connect points J, R, K, and S.

The result is a root-three rectangle (fig. 29).[9]

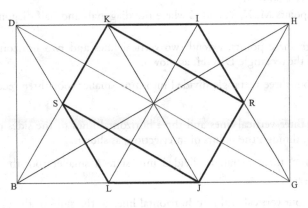

Fig. 29

How to divide a root-three rectangle into smaller root-three rectangles of equal area

- Locate the root-three rectangle (DBGH) of sides 1 and √3.
- Draw the diagonals BH and DG.
- Locate the midpoints (T, U, V, and W) of the root-three rectangle.

- Connect midpoints T and V, then midpoints U and W.

The result is a root-three rectangle divided into four smaller root-three rectangles of equal area (fig. 30).

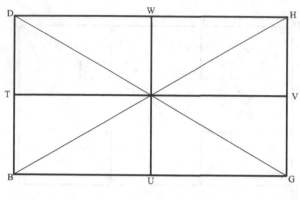

Fig. 30

- From midpoint T, draw the half diagonals TG and TH through the root-three rectangle, as shown.
- From midpoint V, draw the half diagonals VD and VB through the root-three rectangle, as shown.
- From midpoint U draw the half diagonals UH and UD through the root-three rectangle, as shown.
- From midpoint W, draw the half diagonals WB and WG through the root-three rectangle, as shown.
- Locate points M, X, Y, and Z where the diagonals and half diagonals intersect, as shown.
- Through these points, extend two vertical lines and two horizontal lines to the sides of the rectangle DBGH, as shown.

The result is a root-three rectangle divided into nine smaller root-three rectangles of equal area (fig. 31).

- Extend three vertical lines and three horizontal lines to the sides of the rectangle DBGH, through the points of intersection, as shown.

The result is a root-three rectangle divided into sixteen smaller root-three rectangles of equal area (fig. 32).

- Extend four vertical and four horizontal lines to the sides of the original rectangle through the points of intersection, as shown.

The result is a root-three rectangle divided into twenty-five smaller root-three rectangles of equal area (fig. 33).[10]

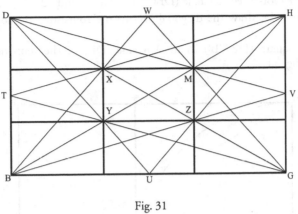

Fig. 31

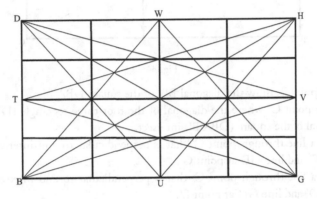

Fig. 32

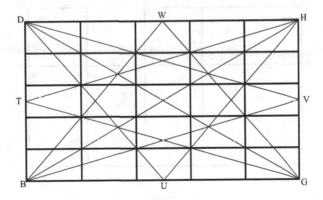

Fig. 33

How to divide a root-three rectangle into patterns of proportional areas

- Locate the root-three rectangle (DBGH) of sides 1 and √3.
- From point D, draw the diagonal DG of the root-three rectangle (DBGH), as shown.
- Apply a square (DBA'B') to the left side (DB) of the root-three rectangle, as shown (fig. 34).

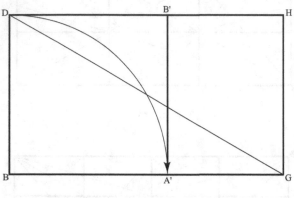

Fig. 34

- From point B, draw the diagonal BB' of the square DBA'B'
- Locate point C' where the diagonal of the root-three rectangle (DBGH) and the diagonal of the square (DBA'B') intersect.
- Draw a line through point C' that is perpendicular to and intersects line BG at point F' and line HD at point G'.
- Draw a line through point C' that is perpendicular to and intersects line DB at point D' and line GH at point E'.

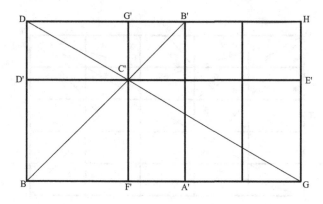

Fig. 35

When a square (DBA'B') is "applied" on the short side of a root-three rectangle (DBGH), it falls short and is elliptic. The square (DBA'B') is composed of two squares and two root-three rectangles. The excess area (A'GHB') is composed of two squares and two root-three rectangles (fig. 35).

- From point H, draw the diagonal HB of the root-three rectangle (DBGH), as shown.
- Apply a square (GHI'H') to the right side (GH) of the root-three rectangle, as shown.
- From point G, draw the diagonal GI' of the square (GHI'H').

When squares (DBA'B' and GHI'H') are "applied" on both short sides of a root-three rectangle (DBGH), they overlap (fig. 36).

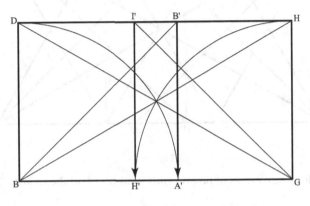

Fig. 36

- From point D, draw the diagonal DA' of the square DBA'B'.
- From point H, draw the diagonal HH' of the square GHI'H'.

The various diagonals reveal a pattern of six squares and five root-three rectangles (fig. 37).

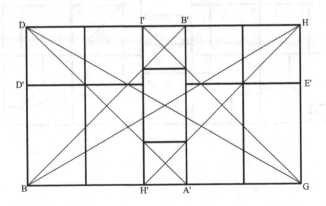

Fig. 37

Design applications

- Locate the root-three rectangle (DBGH) of sides 1 and √3.
- Construct a diagonal grid composed of:
  - the diagonals of the major root-three rectangle (DBGH)
  - the diagonals of the three reciprocals (BLKD, LJIK, and JGHI)
  - the diagonals of the two squares (DBA′B′ and GHI′H′). (See figure 38.)

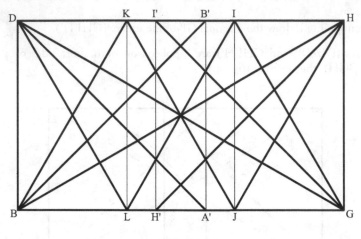

Fig. 38

Figures 39-41 illustrate patterns of proportional areas based on the ratio 1:√3.

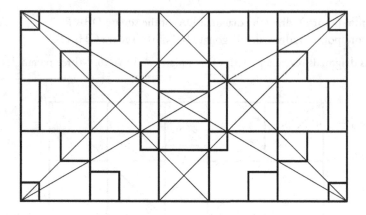

Fig. 39

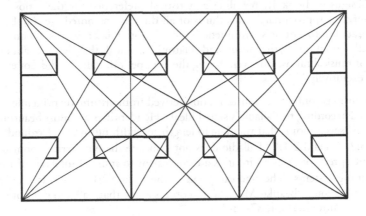

Fig. 40

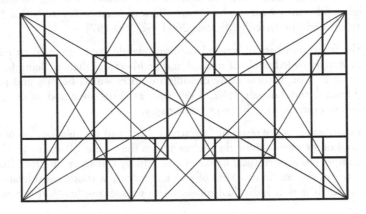

Fig. 41

Applications: A Palladian Villa and Palazzo

That harmonic proportions grace the architecture of Andrea Palladio is without dispute, having been rigorously examined and researched. In the twentieth century, Rudolf Wittkower proposed the theory in *Architectural Principles in the Age of Humanism*, analyzing the role of musical ratios in eight Palladian designs published in *I quattro libri dell'architettura* (*The Four Books on Architecture*). This arithmetic system of proportion was well known in Palladio's day through the treatise *Della proportione, et proportionalità* (*On Ratio and Proportion*), published in 1573 by friend and colleague Silvio Belli, and through Daniele Barbaro's 1556 edition of Vitruvius, which Palladio illustrated [Wittkower 1971,102 137, Ackerman 1966, 160-162].

Following Wittkower, Deborah Howard and Malcolm Longair analyzed all forty-four Palladian designs in Book II, revealing a statistical preference for the ratios of standard musical intervals. Approximately two-thirds of all dimensions noted are numbers that can be incorporated into music's arithmetic ratios.[11] Some later works commissioned by patrons well versed in Pythagorean and Platonic musical theories are almost entirely composed of musical ratios. But even during the later period, Palladio did not employ these techniques exclusively.[12]

Not all dimensions conform to musical ratios derived from arithmetic principles, nor are all rooms shaped according to Palladio's stated ideal. He identifies as "most beautiful and well proportioned" seven rooms that translate to length-to-width ratios of 1:1 (round or square), $\sqrt{2}$:1, 4:3, 3:2, 5:3 and 2:1.[13] Palladio does not associate these preferred room shapes with a musical interpretation, but their ratios are equivalent to musical intervals of Pythagorean and Just tuning systems. The one exception is the ratio $\sqrt{2}$:1, which is a ratio of equal temperament, although Branko Mitrović and others argue that Palladio did not promote its musical value [Mitrović 2004, 87-92].[14]

Mitrović observes that of 153 length-to-width ratios noted in Book II, 89 or 55% correspond to Palladio's stated preferences, including three whole number approximations of the incommensurable ratio $\sqrt{2}$:1. An analysis of Palladio's executed buildings, based largely on eighteenth-century surveys by Ottavio Bertotti Scamozzi, reveals that 54 or 57% of 95 ratios studied are "preferred" [Mitrović 2004, 64, 190-197].[15]

Employing the finite numbers of musical harmony would have appealed to patrons like Barbaro, who were knowledgeable of musical and architectural theory. Being for the most part multiples of 2, 3 or 5, these easily divisible numbers would be easy and practical to work with [Howard and Longair 1982, 136]. And as Wittkower and others note, such ratios are suited to expressions of harmony and beauty.

Palladio says that beauty is the result of "a graceful shape and the relationship of the whole to the parts, and of the parts among themselves and to the whole" [Palladio 1997: I, 7 (7)]. The whole number ratios of musical harmony can achieve this to great effect. But Palladio commonly specifies the inside measures of isolated rooms, without consideration of wall thickness. The result is that individual rooms relate to one another, but not to the total plan.

With nearly half of all length-to-width ratios unaccounted for, Mitrović considers the possible use of other ratios known in the Renaissance. Among the published plans in Book II, he identifies six ratios composed of whole numbers that closely approximate the incommensurable ratio $\sqrt{3}$:1 (1.7320...: 1). These include the large corner rooms of Villa Almerico ("La Rotunda"), which measure 26 x 15 *piedi* and translate to a ratio of 1.7333....: 1.[16] In addition, Mitrović observes approximate root-three ratios in four Bertotti Scamozzi surveys of executed buildings.[17]

Is it possible that root-three ratios functioned dynamically, not merely in the proportions of single rooms, but throughout the plan, inclusive of wall thickness? Two Palladian designs in Book II are worthy of consideration. One—Palazzo della Torre in Verona—was conceived circa 1565, but never executed [Palladio 1997: II, 76 (154)]. The other—Villa Mocenigo, Marocco in Treviso—was designed circa 1559-1562 and partially built soon thereafter, but was destroyed in the early nineteenth century.[18] The outline of each plan approximates a figure of $3\sqrt{3}{:}4$, formed by overlapping identical root-three rectangles. Root-three symmetry persists through multiple divisions, locating the outlines of individual rooms within (figs. 42 and 43).[19]

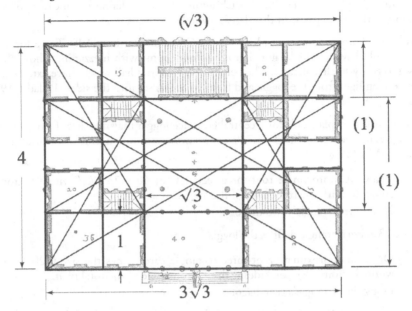

Fig. 42. Palazzo della Torre, Verona (project). Palladio 1570, Book II, 76, with overlay

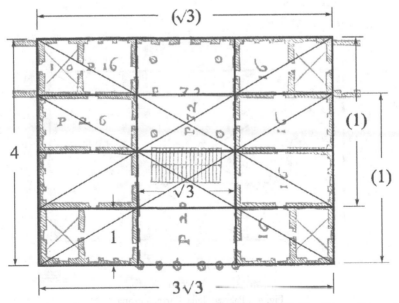

Fig. 43. Villa Mocenigo, Marocco. Palladio 1570, Book II, 54, with overlay

Conventional scholarship disputes the use of incommensurable ratios in Renaissance architecture, in favor of modules comprised of finite numbers.[20] When incommensurable values are recognized, they are approximated in whole numbers, as Mitrović observes in Villa Rotunda and elsewhere. But if such approximations define the measure of a single room, might they not appear in the plan as a whole?

Palazzo della Torre, commissioned by Count Giovanni Battista Della Torre for a site in Verona, would have featured vaulted ground-floor rooms with mezzanines above the small rooms, served by small staircases. The height of the central hall was to have extended to the roof, receiving light from the loggia and from windows set into the sides [Palladio 1997: II, 76 (154)].

Palladio's published plan calls for a central entrance loggia that measures forty *piedi* across. A rectangle in the ratio 40:23 (1.739...: 1) deviates from a true root-three rectangle (1.732...: 1) by 0.4%.

- Draw an approximate root-three rectangle of 40 x 23 *piedi* whose bottom edge coincides with the front edge of the loggia.
- Draw the rectangle's two diagonals.

The 40 x 23 *piedi* rectangle defines the loggia.[21]

- Draw an approximate root-three rectangle three times the size (120 x 69 *piedi*) whose bottom edge coincides with the front edge of the total plan.
- Draw the rectangle's two diagonals.

The top edge of the 120 x 69 *piedi* rectangle locates the inside walls of the rooms along the back (fig. 44).

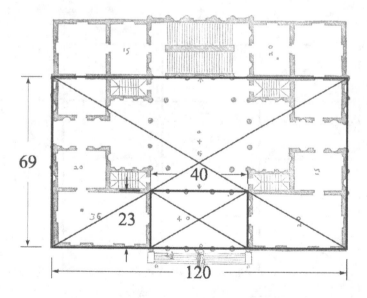

Fig. 44. Palazzo della Torre, Verona

- Extend the top edge of the 40 x 23 *piedi* rectangle to the right and left sides of the 120 x 69 *piedi* rectangle.
- Draw a new 120 x 69 *piedi* rectangle, as shown.
- Draw the rectangle's two diagonals.

The bottom edge of the new 120 x 69 *piedi* rectangle locates the inside walls of the rooms along the front (fig. 45).

- Draw a rectangle that encloses both 120 x 69 *piedi* rectangles.

The result is a 120 x 92 *piedi* rectangle (1.3043....: 1) that locates the outer walls of the total plan. The 120 x 92 *piedi* rectangle approximates a rectangle of 3√3:4 (1.2990...:1). (See figure 45.)

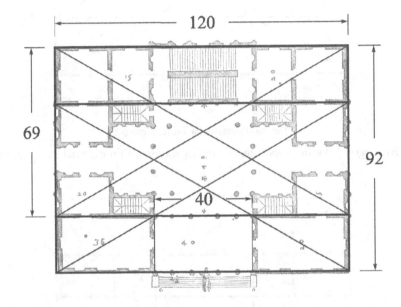

Fig. 45. Palazzo della Torre, Verona

- Locate the four points where the diagonals and horizontal sides of the 120 x 69 *piedi* rectangles intersect.
- Draw diagonals from these four points to the four corners of the 120 x 92 *piedi* rectangle, as shown.

The four new diagonals locate approximate reciprocals (69 x 40 *piedi* or 1.725:1) of the major 120 x 69 *piedi* rectangles (fig. 46).[22]

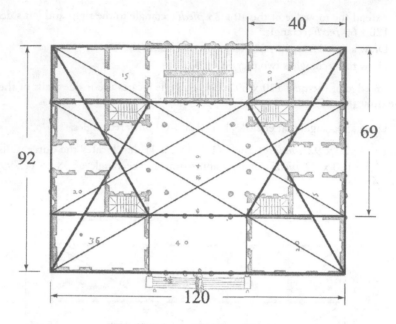

Fig. 46. Palazzo della Torre, Verona

The twelve diagonals locate proportionally smaller rooms within the total plan (fig. 47).

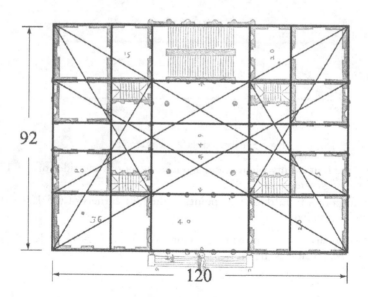

Fig. 47. Palazzo della Torre, Verona

Palladio's collaboration with Venetian nobleman Leonard Mocenigo began in the early 1550s and produced a number of villas, among them Villa Mocenigo in the town of

Marocco on the road from Venice to Treviso [Battilotti 1990, 109]. The published design features a double hexastyle loggia surmounted by a pediment. The lower loggia is Ionic. The upper loggia is Corinthian. Inside, a pair of freestanding staircases ascends in opposite directions, separating the loggia from a hall with four corner columns at the back. Stables, porticos and other amenities comprise a rear court toward the back of the central block [Palladio 1997: II, 54 (132)].

Palladio's published plan calls for an entrance loggia and hall that measure 32 *piedi* across. A rectangle in the ratio 32:18.5 (1.7297…: 1) deviates from a true root-three rectangle (1.7320…: 1) by 0.13%.

- Draw an approximate root-three rectangle of 32 x 18.5 *piedi* whose bottom edge coincides with the front edge of the loggia.
- Draw the rectangle's two diagonals.

The 32 x 18.5 *piedi* rectangle defines the front edge of the loggia and the width of adjacent rooms along the front.[23]

- Draw an approximate root-three rectangle three times the size (96 x 55.5 *piedi*) whose bottom edge coincides with the front edge of the total plan.
- Draw the rectangle's two diagonals.

The top edge of the 96 x 55.5 *piedi* rectangle locates the inside walls of the rooms along the back (fig. 48).

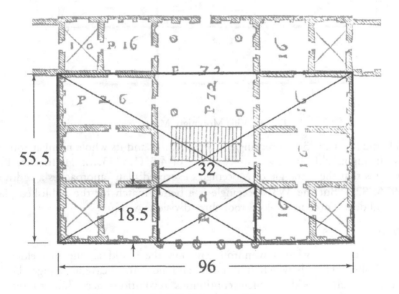

Fig. 48. Villa Mocenigo, Marocco

- Extend the top edge of the 32 x 18.5 *piedi* rectangle to the right and left sides of the 96 x 55.5 *piedi* rectangle.
- Draw a new 96 x 55.5 *piedi* rectangle, as shown.

- Draw the rectangle's two diagonals.

The bottom edge of the new 96 x 55.5 *piedi* rectangle locates the inside walls of the rooms along the front.

- Draw a rectangle that encloses both 96 x 55.5 *piedi* rectangles.

The result is a 96 x 74 *piedi* rectangle (1.2972....: 1) that locates the outer walls of the total plan.

The 96 x 74 *piedi* rectangle approximates a rectangle of 3√3:4 (1.2990...:1). The four diagonals locate proportionally smaller rooms within the total plan (fig. 49).

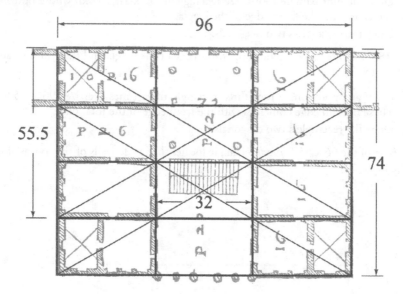

Fig. 49. Villa Mocenigo, Marocco

One could argue that the 3√3:4 rectangle (1.2990....: 1) and its whole number equivalents are barely distinguishable from a simpler figure of 4:3 (1.3333....: 1). In fact, Bertotti Scamozzi notes that the rectangle of Mocenigo's executed plan "approaches a width and a third" [1976, 82]. But the 3√3:4 rectangle is a better match for the published plans of Mocenigo and della Torre and reflects root-three divisions within.[24]

Conclusion

> ...if we consider what a wondrous creation the world is, the marvelous embellishments with which it is filled, and how the heavens change the seasons of the world by their continuous revolutions according to the demands of nature and how they maintain themselves by the sweetest harmony of their measured movements, we cannot doubt that, since these small temples which we build must be similar to this vast one which He, with boundless generosity, perfected with but a word of command, we are bound to include in them all the embellishments we can, and build them in

such a way and with such proportions that together all the parts convey to
the eyes of onlookers a sweet harmony...

–Palladio, *I quattro libri* [1997: IV, 3 (213)]

The geometric studies presented here are based on Palladio's original woodcuts from the 1570 Italian edition of *I quattro libri* [Palladio 1980, II, 54, 76]. The drawings are fairly conceptual and abstract, hardly ideal for precise geometric analysis.[25] But they provide clues about the ideal measures and proportions conceived by Palladio for them, before these were modified by constraints of site or execution.[26]

Because there is no documentary evidence in the form of working drawings or theoretical writings that attest to the intentional use of incommensurable proportions, it is possible these relationships are due to chance. But the elegant way in which root-three ratios achieve spatial harmony through multiple divisions is compelling.

Like Palladio's conception of the heavens, dynamic symmetry responds to our innate desire for harmony to permeate the world. When applied even in an approximate way to commensurable figures, as illustrated here, it can unite a diversity of elements and invoke delight.

Notes

1. "Dynamic symmetry" appears in root rectangles based on square root proportions. The edge lengths of such rectangles are incommensurable and cannot divide into one another. But a square constructed on the long side of the rectangle can be expressed in whole numbers, relative to a square constructed on the shorter side [Fletcher 2007, 327-328; Hambidge 1960, 22–24; 1967, 17–18]. See [Fletcher 2007, 328-334] to construct an expanding series of root rectangles from a square. The diagonal of the preceding square or rectangle equals the long side of the succeeding four-sided figure. The short side of each root rectangle is 1. The long sides progress in the series $\sqrt{2}$, $\sqrt{3}$, $\sqrt{4}$, $\sqrt{5}$, $\sqrt{6}$... . See [Fletcher 2007, 344-346] to construct diminishing root rectangles within a square. The long side of each root rectangle is 1. The short sides progress in the series $1/\sqrt{2}$, $1/\sqrt{3}$, $1/\sqrt{4}$, $1/\sqrt{5}$... . See [Fletcher 2007] for more on dynamic symmetry and its fundamental components.

2. The calculation of the diagonal BC is based on the Pythagorean Theorem. Triangle CDB is a right triangle. $CD^2 + DB^2 = BC^2$ [$1^2 + 1^2 = 2$]. Thus, $BC = \sqrt{2}$. The diagonal of a square of side 1 is equal to $\sqrt{2}$. The diagonal that appears in figure 5 can be calculated in similar fashion. See [Fletcher 2005b: 44-45] for more on the Pythagorean Theorem.

3. The reciprocal of a major rectangle is a figure similar in shape but smaller in size, such that the short side of the major rectangle equals the long side of the reciprocal [Hambidge 1967, 30, 131].
    A root-two rectangle is in the ratio 1:1.4142.... Its reciprocal ($\sqrt{2}/2$:1) is in the ratio 0.7071... :1.
    A root-three rectangle is in the ratio 1:1.732.... Its reciprocal ($\sqrt{3}/3$:1) is in the ratio 0.5773... :1.
    A root-four rectangle is in the ratio 1:2.0. Its reciprocal ($\sqrt{4}/4$:1) is in the ratio 0.5: 1.
    A root-five rectangle is in the ratio 1:2.236... . Its reciprocal ($\sqrt{5}/5$:1) is in the ratio 0.4472...:1. See [Fletcher 2007, 336-343] and [Hambidge 1967, 31–38] for more on these element.

4. Hambidge identifies three ways to apply one area to another—less than, equal to, or in excess of the other. If the applied area is less, it is elliptic; if equal, it is parabolic; and if in excess, it is hyperbolic. See [Fletcher 2008, 160] and [Hambidge 1920, 19-20; 1960, 35] for more on the application of square areas.

5.	In Hambidge's system, each rectangle has a reciprocal and each rectangle and reciprocal have complementary areas. The complementary area is the area that remains when a rectangle is produced within a unit square. If the rectangle exhibits properties of dynamic symmetry, its complement will also. See [Fletcher 2007, 347-354] and [Hambidge 1967, 128].

6.	Rectangles are similar if their corresponding angles are equal and their corresponding sides are in proportion. Similar rectangles share common diagonals.

7.	See [Hambidge 1967, 48-50; Schneider 2006, 109-117].

8.	The radius vector is the variable line segment drawn to a curve or spiral from a fixed point of origin (the pole or eye) [Simpson 1989]; see also [Fletcher 2004, 105].

9.	For more on the root-three rectangle and its relationship to the vesica piscis and the hexagon, see [Fletcher 2004, 102–105; and 2005a, 153–157].

10.	See [Schneider 2006, 30-32, 34] for similar constructions.

11.	Of 365 numbers appearing in the plans of Book II, 239 or 65.5% are harmonic, whereas a random sample should produce less than half [Howard and Longair 1982, 116, 122, 136]. Palladio's designs are measured by the Vicentine foot (*piede vicentino*), which divides into twelve inches and four minutes [Palladio 1997: II, 4 (79)].

12.	Comprehensive harmonic plans, such as Palladio's Villa Barbaro at Maser, postdate Barbaro's 1556 edition of Vitruvius [Howard and Longair 1982, 136].

13.	"There are seven types of room that are the most beautiful and well proportioned and turn out better: they can be made circular, though these are rare; or square; or their length will equal the diagonal of the square of the breadth; or a square and a third; or a square and a half; or a square and two-thirds; or two squares." [Palladio 1997: I, 52 (57)]. Similar lists are proposed by Vitruvius [1999: VI, iii, (79)], Sebastiano Serlio [1996: I, 15r (30)] and other Renaissance theorists.

14.	In a compelling interpretation, Lionel March observes that between the numbers 1 and 2 are four proportional means that produce the remaining numbers in the series: the arithmetic mean (3/2); the geometric mean ($\sqrt{2}/1$); the harmonic mean (4/3); and the contraharmonic mean (5/3). All four are noted in Belli's treatise [March, "Forward" in Belli 2002, p. 11].

15.	For a discussion of the relative accuracy of Bertotti Scamozzi's surveys, see [Mitrović 2004, 30].

16.	[Palladio 1997: II, 19 (95).] Additional length-to-width ratios that approximate the ratio $\sqrt{3}$:1 include: the ratio 19:11 (1.7272...: 1) in Palazzo della Torre, Verona [II, 11]; the ratio 34.5:20 (1.725:1) in Palazzo Thiene, Vicenza [II, 13] and in an unnamed Venetian palace [II, 72]; the ratio 12:7 (1.7142...: 1) in Palazzo Barbarano, Vicenza [II, 22]; the ratio 28:16 (1.75:1) in Villa Pisani, Montagnana [II, 52] and 14:8 (1.75:1) in Palazzo Trissino, Vicenza [II, 74]; and the ratio 30:17 (1.7647...: 1) in Villa Poiana, Poiana Maggiore [II, 58] [Mitrović 2004, 65, 197].

17.	These include: the ratio 1.718...: 1 (26' 6" x 15' 5") in Villa Cornaro, Piombino Dese [II, 53]; the ratio 1.775:1 (35' 6" x 20') in Palazzo Thiene, Vicenza [II, 13]; the ratio 1.714...: 1 (30' x 17' 6") in Villa Pisani, Bagnolo [II, 47]; and the ratio 1.711...: 1 (29' 8" x 17' 4") in Villa Poiana, Poiana Maggiore [II, 58] [Mitrović 2004, 197]. For a discussion of other incommensurable ratios, see [Mitrović 2004, 65-70].

18.	[Palladio 1997: II, 54 (132).] See [Puppi 1986, 171].

19.	Base images for figs. 42-49 are from [Palladio 1980, 54, 76] with geometric overlays by the author. Root-three symmetry does not extend to the inside dimensions of individual rooms, which tend to reflect "preferred" room shapes and harmonic numbers. In *I quattro libri*, the rooms at Palazzo della Torre [II, 76] measure 36 x 20 *piedi* (1.8:1), 20 x 15 *piedi* (4:3), and 20 x 20 *piedi* (1:1). All four dimensions noted—15, 20, 36, 40—are harmonic numbers. Two length-to-width ratios—20:15 (4:3) and 20:20 (4:3)— are "preferred" and one—36:20 (1.8:1) is not. No rooms contain inside measures in approximate root-three ratio. The rooms at Villa Mocenigo, Marocco [II, 54] measure 16 x 10 *piedi* (1.6:1), 16 x 16 *piedi* (1:1), and 26 x 16 *piedi* (1.625:1), and 32 x 32 *piedi* (1:1). Of five dimensions noted—10, 16, 20, 26 and 32—all but one (26) are harmonic numbers. Two square rooms proportioned to 16:16 and

32:32 (1:1) are "preferred." Two rooms proportioned to 16:10 (1.6:1) and 26:16 (1.625:1) are not. No rooms contain inside measures in approximate root-three ratio [Howard and Longair 1982, 139-143, Mitrović 2004, 192-194]. In Bertotti Scamozzi's survey of the executed Villa Mocenigo, the 16' square room measures 17' 8" x 16' 4", the 26' length of the larger rooms measures 26' 11", and the 10' width of the small rooms measures 8'5" [Bertotti-Scamozzi 1976, 83].

20. Mitrović cites Vitruvius, Barbaro and Palladio to argue that commensurable ratios alone were used to create Renaissance designs, including whole number approximations of incommensurable ratios such as 1:√3 [Mitrović 2004, 66].

21. The 40 *piedi* measure is exclusive of wall thickness. The 23 *piedi* measure in inclusive of wall thickness.

22. The diagonals of a true root-three rectangle and its reciprocal intersect at 90.0°. In this approximate construction, they intersect at 89.8° and 90.2°, a deviation of 0.2%.

23. The 32 *piedi* measure is exclusive of wall thickness. The 18.5 *piedi* measure in inclusive of wall thickness.

24. The outline of Bertotti Scamozzi's plan of Mocenigo falls between 3√3:4 and 4:3 rectangles. The 4:3 rectangle is a better match for the published plan of Villa Valmarana at Lisiera [Palladio 1980, II, 59].

25. [Mitrović 2004, 34-35; Wittkower 1971, 128.] Though more than a decade in preparation, Palladio's manuscript was hastily assembled at the time of publication. Drawings are executed with varying levels of precision, often lacking sufficient regard for practical considerations to function as working drawings. Sometimes the drawings contradict the text or contain elements impossible to realize in the physical world. For example, room dimensions can discount the thickness of walls, as in the plan for Villa Mocenigo, where two rooms with inside measurements of 10 x 16 *piedi* and 16 x 16 *piedi* occupy the same space as a single room of 16 x 26 *piedi*. Discrepancies between published dimensions and executed plans can sometimes be attributed to printers' errors and sometimes to revisions made during the design process. For a full discussion, see [Howard and Longair 1982, 117-118].

26. Sometimes, proportions expressed on paper do not survive the building process. In one instance, published and executed plans possibly express different proportional concepts. March and Wittkower demonstrate 3:4:5 triangles and elaborate musical harmonies in the published plans of Villa Emo, while the author observes golden mean ratios throughout the executed plan. See [Fletcher 2000; Fletcher 2001; March 2001; Wittkower 1971, 131].

References

ACKERMAN, James S. 1966. *Palladio*. New York: Penguin.

BATTILOTTI, Donata. 1990. *The Villas of Palladio*. Richard Sadleir, trans. Milan: Electa.

BELLI, Silvio. 2002. *On Ratio and Proportion: The Common Properties of Quantity*. Trans. and comm. Stephen R. Wassell and Kim Williams. Fucecchio, Florence: Kim Willams Books.

BERTOTTI SCAMOZZI, Ottavio. 1976. *The Buildings and the Designs of Andrea Palladio*. 1776. Howard Burns, trans. Trent: Editrice La Roccia.

EUCLID. 1956. *The Thirteen Books of Euclid's Elements*. Thomas L. Heath, ed. and trans. Vols. I–III. New York: Dover.

FLETCHER, Rachel. 2000. Golden Proportions in a Great House: Palladio's Villa Emo. Pp. 73-85 in *Nexus III: Architecture and Mathematics*. Kim Williams, ed. Pisa: Pacini Editore.

———. 2001. Palladio's Villa Emo: The Golden Proportion Hypothesis Defended. *Nexus Network Journal* **3**, 2 (Summer-Autumn 2001): 105-112.

———. 2004. Musings on the Vesica Piscis. *Nexus Network Journal* **6**, 2 (Autumn 2004): 95–110.

———. 2005a. Six + One. *Nexus Network Journal* **7**, 1 (Spring 2005): 141–160.

———. 2005b. The Square. *Nexus Network Journal* **7**, 2 (Autumn 2005): 35–70.

———. 2007. Dynamic Root Rectangles. Part One: The Fundamentals. *Nexus Network Journal* **9**, 2 (Spring 2007): 327-361.

―――. 2008. Dynamic Root Rectangles. Part Two: Root-Two Rectangles and Design Applications. *Nexus Network Journal* **10**, 1 (Autumn 2008): 149-178.

HAMBIDGE, Jay. 1920. *Dynamic Symmetry: The Greek Vase*. New Haven: Yale University Press.

―――. 1960. *Practical Applications of Dynamic Symmetry*. 1932. Rpt. New York: Devin-Adair.

―――. 1967. *The Elements of Dynamic Symmetry*. 1926. Rpt. New York: Dover.

HOWARD, Deborah and Malcolm LONGAIR. 1982. Harmonic Proportion and Palladio's *Quattro Libri*. *Journal of the Society of Architectural Historians* **41**, 2 (May 1982): 116-143.

MARCH, L. 2001. Palladio's Villa Emo: The Golden Proportion Hypothesis   Rebutted,   N*exus Network Journal* **3**, 2 (Summer-Autumn 2001): 85-104.

MITROVIĆ, Branko. 2004. *Learning from Palladio*. New York: W. W. Norton & Company.

OED ONLINE. Oxford: Oxford University Press. 2004. http://www.oed.com/

PALLADIO, Andrea. 1965. *The Four Books of Architecture*. Intro. Adolf N. Placzek. London: Isaac Ware, 1738. Facsimile, New York: Dover.

―――. 1980. *I quattro libri dell'architettura*. Venice: Dominico de'Franceschi, 1570. Facsimile, Milan: Ulrico Hoepli Editore.

―――. 1997. *The Four Books on Architecture*. Trans. Robert Tavernor and Richard Schofield. Venice: Dominico de'Franceschi, 1570. Reprint, Cambridge: The MIT Press.

PUPPI, Lionello. 1986. *Andrea Palladio: The Complete Works*. Trans. Pearl Sanders. New York: Electa/Rizzoli.

ROBISON, Elwin C. 1988-89. Structural Implications in Palladio's Use of Harmonic Proportions. *Annali d'Architettura* **10/11** (1988-99): 175-182.

SCHNEIDER, Michael S. 2006. *Dynamic Rectangles: Explore Harmony in Mathematics and Art*. Constructing the Universe Activity Books. Vol. IV. Self-published manuscript.

SERLIO, Sebastiano. 1996. *Sebastiano Serlio on Architecture*. Vol. I. Books I-V of *Tutte l'Opere D'Architettura et Prospetiva*. Trans. Vaughan Hart and Peter Hicks. New Haven: Yale University Press.

SIMPSON, John and Edmund WEINER, eds. 1989. *The Oxford English Dictionary*. 2nd ed.

VITRUVIUS. 1999. *Ten Books on Architecture*. Ingrid D. Rowland, trans. Cambridge: Cambridge University Press.

WITTKOWER, Rudolf. 1971. *Architectural Principles in the Age of Humanism*. New York: W. W. Norton & Company.

About the Geometer

Rachel Fletcher is a theatre designer and geometer living in Massachusetts, with degrees from Hofstra University, SUNY Albany and Humboldt State University. She is the creator/curator of two museum exhibits on geometry, "Infinite Measure" and "Design by Nature". She is the co-curator of the exhibit "Harmony by Design: The Golden Mean" and author of the exhibit catalog. In conjunction with these exhibits, which have traveled to Chicago, Washington, and New York, she teaches geometry and proportion to design practitioners. She is an adjunct professor at the New York School of Interior Design. Her essays have appeared in numerous books and journals, including *Design Spirit, Parabola,* and *The Power of Place*. She is the founding director of the Housatonic River Walk in Great Barrington, Massachusetts, and is currently directing the creation of an African American Heritage Trail in the Upper Housatonic Valley of Connecticut and Massachusetts.

Book Review

Branko Mitrović and Stephen R. Wassell (eds.)

Andrea Palladio: The Villa Cornaro in Piombino Dese

New York: Acanthus Press, 2007.

Reviewed by Kim Williams

Via Cavour, 8
10123 Turin (Torino) ITALY
kwb@kimwilliamsbooks.com

Keywords: Renaissance architecture, Palladio, surveying techniques, proportional analysis,

I cannot claim to be the most objective reviewer of *Andrea Palladio: The Villa Cornaro in Piombino Dese,* edited by Branko Mitrović and Stephen R. Wassell, with contributions by Tim Ross and Melanie Bourke. I have known both of the editors for a good while and have spoken to them at length about the Villa Cornaro and Andrea Palladio, and even participated (in a very minor way, and principally by bringing along someone who did the dirty work) in the survey campaign. Some of my most pleasant hours have been spent with Sally and Carl Gable in "their" villa in Piombino Dese. On the other hand, what I lack in objectivity, the book amply makes up for in objectivity of its own, just one of the many respects in which this is not just another book about Palladio.

While most books set forth a theory, and use data to support it (in the case of architecture, measurements), this particular book on the Villa Cornaro does, of course, set forth theories, but these are not supported by selected data, but rather what is furnished is an abundance of data. One very great contribution that this book will make to Palladio scholarship is in providing a body of information that future Palladio scholars can turn to in elaborating and verifying their own theories, without being forced to agree with those set forth by Mitrović and Wassell. This is objectivity indeed, and more, generosity.

The essays

Mitrović's essay, "Designing the Villa Cornaro", begins with an overview of the current status of scholarship on Palladio's design theory, and then proceeds to contrast Mitrović's own theory with previous ones. Mitrović explains his principle of the "condition of concordance of heights", which he calls the CCH rule, through which the interrelatedness of Palladio's volumes is made clear. The information gleaned from the survey of the Villa Cornaro has allowed him to construct an in-depth proof of a hypothesis regarding Palladio's design process applied to a specific example that he had presented with broader strokes in his earlier publication, *Learning from Palladio* (W.W. Norton, 2004). His essay also provides the context for the detailed analysis of the projections of the attic base in the essay that follows.

When at the beginning of his chapter "The Piombino Dese *Piede*", Wassell writes, "I can state that the basic unit of measurement used in the construction of the Villa Cornaro

… is equivalent to 38.4 cm…" there is justly the ring of triumph. This is a simple datum, obtained after much hard work, a gold nugget found after sifting through lots of sand and grit. The guidelines for treating the data, including the philosophy regarding the treatment of the possible kinds of error, are explained in detail in the Introduction by Mitrović and Wassell.

Consider, for example, Wassell's particular study of the column bases on pp. 47-49. In order to see if what Palladio indicates in the *Quattro libri* for the projections of the attic base elements is put into practice in the Villa Cornaro, Wassell analyzes Palladio's illustration to produce the rules, next converts the rules to a measurable standard in terms of his survey, and then completes his investigation. In an appendix he supplies fourteen tables with a total of 448 raw measurements and 90 means and medians (this is data crunching). This is a massive bit of work to examine a minute architecture detail. It is justified because Palladio himself laid out such detailed proportions in his treatise: this is exactly how to see if Palladio followed his own specifications. For although a large part of the appeal and popularity of Palladio's work can be attributed to its accessibility and the visual clarity of his forms, underlying these are Palladio's formidable theories about how the individual elements relate to each other. So it should come as no surprise that the means for verifying them are formidable as well. In essence, the line drawings in the plates at the end of this book are deceptively simple, in the same way that Palladio's architecture is. Both drawings and architecture are clean and clear; both are based on an enormous quantity of information.

Other specific elements are described in the final two essays. Tim Ross compares the principal doors of the Villa Cornaro with the proportions set forth in the *Quattro libri*. I thought that the most interesting part of this discussion concerned "contractions", that is, the narrowing of the width of the door at its top, a concept similar to the entasis of a column. The essay by Melanie Bourke examines the two identical internal staircases that flank the back porch of the villa. Although these stairs are tucked away, and not as Palladio recommended, "still … obvious and easy to find", and although they are in plain brick and otherwise unadorned, they are particularly pleasant to use

The survey

Another contribution that this book will make is that it will set a standard for future survey projects and their presentation. Clarity has always been a hallmark of Mitrović's scholarly work: he not only tells you what his theory is, but he also explains his reasons for taking a particular approach towards the question. This same clarity is apparent in the Appendix, where Wassell, Bourke and Ross explain how the survey project was approached and the methodologies used.

My own broad experience with surveying and measuring has taught me that there are two equally arduous parts of the survey campaign: gathering the raw measurements and "crunching the data", that is, turning those raw measurements into a form that is meaningful, useful and accessible. Too much data can be as meaningless as too little.

Another reason that this is an unusual book is that it is rather a hybrid. One sense in which it is hybrid is that it is a collaboration between an architecture historian (Mitrović) and a mathematician (Wassell), although this distinction grossly oversimplifies the domains of knowledge of each, and I apologize to the authors in advance. In another sense, it is a combination of sixteenth-century architectural theory and twenty-first-century technology.

The book's large 30 x 42,5 cm format was necessary in order to accommodate the plates that report the results of the survey. (Not all publishers, myself included, are capable of producing this kind of book, and Barry Cenower of Acanthus Press is certainly to be commended.) But given the nature of the data presented, and the fact that the data was largely produced with the aid of a computer, one marvels that it is a book at all – the data almost cries out for a digital support. For instance, figs. 51 and 52 on p. 68 show the point cloud and the wireframe model of the Villa Cornaro, two very recent conceptual constructions. (Given the book's large format, it is too bad that these two figures were not represented at a larger scale.) These make it clear, however, that the original data produced was a set of Cartesian coordinates related to some origin (0,0,0) from which the linear dimensions were produced. From the original set of coordinates, countless other dimensions could be had; it might even be possible to do much more modern kind of analysis, such as box-counting to determine fractal dimension. On the other hand, of course it is a book, because it is Palladio, for whom print was arguably as much of a medium for propagations of his ideas as his buildings were, and of course the coordinates are not given, for the linear dimensions are the only ones that Palladio himself would have worked with.

Naturally, a survey campaign of this scope makes perfect sense for an analysis of Palladio. Would it make as much sense, or provide, in the final analysis, the same quantity of useful information for a much later architect such as Louis Kahn? Although Kahn left no written texts on his preferred proportions or a system of proportions, he did leave traces in his buildings that have been sought out and convincingly presented. The question isn't even remotely relevant for a contemporary architect such as Gehry, who does not take formal proportions into consideration, and so for whose architecture accurate measurements add little or nothing to what one knows about the building.

What I perceived in this book is not so much the desire to undertake an arcane exercise in architectural history as a pressing need to set down what has been learned for posterity *so we won't forget*. Throughout the centuries interest in proportion has waxed and waned; in some epochs it is of the essence, while in others it is completely neglected. We live in an age when proportion does not much matter. But someday it will again. And when that time comes, the effort that Mitrović and Wassell have put into this survey will bear fruit. The metaphor of architecture as frozen music may be a bit worn, but it is still valid. Palladio knew how to make his proportions sing. When the time is right, this book will help assure that future architects can learn that music once again.

Mitrović and Wassell point out that the survey was done in ten days by four people and add "there is little reason to believe that similar surveys of Palladio's other work could not be done with similarly modest resources" (p. 16). They've neglected to mentions years – if not decades – of careful consideration they have given to Palladio's methods and results. Not all surveyors are so well-prepared, and the final result certainly benefits from this experience.

About the reviewer

Kim Williams is the Editor-in-Chief of the *Nexus Network Journal.*

Book Review

Deborah Howard and Laura Moretti (eds.)

Architettura e Musica nella Venezia del Rinascimento

Milan: Mondadori, 2006.

Reviewed by Sylvie Duvernoy

Via Benozzo Gozzoli, 26
50124 Florence ITALY
sduvernoy@kimwilliamsbooks.com

Keywords: Renaissance architecture, acoustics, Palladio

This publication is the proceedings of an international conference which was held on the Isola San Giorgio in Venice on 8-9 September 2005, organized by both the Fondazione Scuola di San Giorgio and the University of Cambridge, where the two editors of the volume, Professor Deborah Howard and Architect Laura Moretti, teach and work.

As professor Howard explains in her introduction to the proceedings:

> The book represents the first stage in an international, interdisciplinary research project based at the Centre for Acoustic and Musical Research in the Renaissance Architecture (CAMERA) in the Department of History of art, University of Cambridge and funded by the Art and Humanities Research Council of the UK [p. 8-9].

The papers of the book are written either in English or in Italian but all the abstracts are in both languages: the book is therefore accessible to a vast international readership.

During the conference on the Isola San Giorgio, the question of the direct relationship between architecture and music in the Renaissance was approached mainly from the standpoint of acoustics, and the architectural spaces that were studied were the churches and chapels of Venice of the sixteenth century. The field of inquiry presented in the proceedings is therefore narrower than what the general title of the book, *Architettura e Musica nella Venezia del Rinascimento*, might suggest at first sight, but it is indeed interesting and discussed in depth. Churches are multifunctional buildings which have to satisfy several requirements that may be related either to liturgy (mass and sermons) or to festivities (music and choir singing). In addition they are composed of various interior spaces and volumes that interact with one another through openings and wide passages that alter the intrinsic acoustics qualities of each space. The research presented here focuses on the interaction between three contemporary aspects of sixteenth-century Venetian culture: the scientific knowledge of acoustics, church architecture and the musical innovation of choral polyphony.

Although the scholars attending conferences on specific topics are usually already experts in the field that is going to be discussed, two of the first papers of the book are short lessons for neophytes on the bases of acoustics and architectural form, and especially on the state of knowledge of architectural acoustics in the late Renaissance. Those papers are fundamental for the comprehension of the further researches, since they seem to present the common criteria for evaluating the architectural spaces under investigation, and they indeed provide the non-specialist but interested reader with the necessary basic notions, without which their colleagues' efforts would be less accessible. Unfortunately there is no similar introduction outlining elementary information about the musical evolution of the sixteenth century and the elaborate form of choral polyphony introduced in Venice at the instigation of the doge Andrea Gritti, for which every church and chapel had to adapt its interior design, permanently or occasionally.

The architectural spaces that are studied are both ancient churches and newly built monuments. In fact,

> while a number of the city's roughly 150 churches where indeed constructed in the sixteenth century, almost all the buildings dated back to earlier periods and had "survived" into the sixteenth century with or without modifications. Sixteenth-century polyphony was performed in all those churches, regularly or intermittently [D. Bryant, E. Quaranta, F. Trentini, p. 261].

The two main monuments with which scholars deal in their process of research are the two major examples of the two categories: the cathedral of San Marco and the church of San Giorgio Maggiore. The first one illustrates evolution and adaptation, as it is a church whose form evolved through time and whose transformations went on during the Cinquecento when polyphonic choral music first developed. The second one is *the* example of modernity, since it was built in the sixteenth century by Palladio in the very days of musical innovation.

The researches presented in this volume will surely offer interesting clues to scholars involved in the study of relationships between mathematics and music, which is a recurrent topic of discussion in the pages of the *Nexus Network Journal* and at the biennial Nexus conferences. And although the field of research presented in this book is very narrow and specific, it considerably enlarges the discussion about architecture and music, which is too often restricted to the question of harmonic proportions inherited from Antiquity and their numerical applications in the geometric patterns of Renaissance architecture.

About the reviewer

Sylvie Duvernoy is the Book Review Editor for the *Nexus Network Journal*.